# The Nation's Got Talent

This book explores the construction of the idea of the 'talented' student in India and its relationship to the discourse of the 'nation'. It historically situates the evolution of the National Science Talent Search (NSTS) and its subsequent avatar, the National Talent Search Examination (NTSE), with state-sponsored ideas and practices of 'nation-building'. It also delves into how individuals who wrote and cleared the examination inhabit this identity of the 'talented'.

Drawing on policy documents and institutional literature of over 50 years as well as interviews with past winners of the NSTS/NTSE, including a Nobel laureate, this book is a major intervention in the field of South Asian studies, public policy, and education.

**Rachel Philip** is an Assistant Professor in the School of Liberal Arts, IIT Jodhpur. Her work explores the social imagination, construction, and application of discourses like 'talent', 'equality', 'quality', 'interest', etc. through the critical documentation of education interventions. With an interdisciplinary background in English Literature, Sociology, and Education, Dr Philip has previously worked with the ICICI Foundation, Pune, and taught at the Departments of Sociology of the Lady Shri Ram College for Women (LSR) and the Indraprastha College for Women (IPCW), University of Delhi. She has also undertaken fieldwork and implemented a pre-school reading program in Innopolis, Russia.

# The Nation's Got Talent

Education, Experimentation and
Policy Discourses in India

Rachel Philip

Routledge
Taylor & Francis Group

LONDON AND NEW YORK

First published 2023
by Routledge
4 Park Square, Milton Park, Abingdon, Oxon OX14 4RN

and by Routledge
605 Third Avenue, New York, NY 10158

*Routledge is an imprint of the Taylor & Francis Group, an informa business*

© 2023 Rachel Philip

The right of Rachel Philip to be identified as author of this work
has been asserted in accordance with sections 77 and 78
of the Copyright, Designs and Patents Act 1988.

*British Library Cataloguing-in-Publication Data*
A catalogue record for this book is available from the British Library

ISBN: 978-1-032-29092-8 (hbk)
ISBN: 978-1-032-38408-5 (pbk)
ISBN: 978-1-003-34490-2 (ebk)

DOI: 10.4324/9781003344902

Typeset in Sabon
by SPi Technologies India Pvt Ltd (Straive)

*To Riby, with love;*
*You set a ceaseless example of striving harder according*
*to the call of Lord Jesus Christ*

# Contents

# Figures

# Tables

# Acknowledgements

It was on a long drive one night from Mumbai to Pune with my friend and then boss, Mili, in 2009 when I got the first real confidence that I could commit to doing a PhD on a topic that had been close to my heart for many years, i.e. the assessment of students and the different kinds of impact that it had on their lives. In the contemplative quiet that followed our conversation, I happened to look outside my window and the star-studded sky stole my breath. It also vividly reminded me of a conversation in the Bible between God and a man called Abraham who 'against all hope, in hope, believed'. '*The Lord brought forth abroad and said, 'Look now toward heaven and tell the stars, if thou be able to number them, and he said unto [Abraham], So shall thy seed be'*. Though a different sort of seed was at the heart of that conversation, I look back at that moment as one when a seed of an idea took root and later grew to be my doctoral thesis. With the completion of this manuscript, my heart overflows with gratitude to the Lord Jesus Christ, the One who is the beginning and the end, who kept the dream alive through the ups and downs of doing research, starting a family and raising children, the pressures of teaching, a move to a new country, and the return home in the midst of the devastating COVID-19 epidemic.

It was my great privilege and joy to undertake this research under the supervision of Professor Krishna Kumar as one of his last PhD students before his retirement. With his trademark blend of insight and wit, he set the tone for this intellectual journey, reminding me at the very outset, 'This is not your life's greatest work. It is a chance to satisfy your curiosity and learn something new'. I deeply appreciate the combination of guidance and freedom which he provided, along with the constant encouragement to reflect on the relationship between my work and my beliefs. Most of all, his faith in me as a scholar and in this work spilled over post the completion of the PhD and in our subsequent interactions, where he encouraged me to publish this work as a book. During the period when we were in Russia, Frances Kumar's emails helped me rediscover joy in the day-to-day aspects of juggling research and other aspects of family life.

I have also received much assistance from many individuals at the NCERT, including the chief librarian Dr Moortimatee Samantaray, Prof Avtar Singh (who is now retired), Prof Indrani Bhaduri, Mr Lingwal (who is also retired), and Mr Ajay. I am grateful to all the NTSE scholars who participated in this study, sharing their memories and reflections on the programme. Many of them went out of their way to point me in the direction of other scholars as well as took a deep interest in the evolution of this work. Post the publication of a part of this work in *Science and Education*, it was tremendously encouraging to receive emails and messages from other NTSE scholars, commenting on the piece as well as writing about their experiences. The publication of this manuscript also owes much to the great team at Routledge with whom I could collaborate.

I made several significant transitions during the research and post the completion of my PhD, including marriage and motherhood. Through it all, my family gave me the freedom to pursue my dreams and took leaps of faith in supporting me during difficult times. I have been so blessed to have wonderful parents, my father, Dr M. G. Philip, and my mother, Subha, who have stood by me with love, faithful prayers, and the occasional 'Just finish it'. My mother went the extra mile, even helping me in the seemingly never-ending process of the transcription of my interviews. My sister, Remya, also assisted me with transcription as well as proofreading my chapters, seasoning feedback with her characteristic wit.

I am exceptionally blessed to be part of a second family, who unstintingly provided their support and encouragement, even before our marriage and not to mention, after! I will never forget the amazing surprise it was to learn that my father-in-law, Prof Boby Abraham (Former Principal, St. Mary's College), and his friends used to conduct a coaching camp for the NTSE in Kottayam in the early 1980s. Their stories provided an interesting counterpoint to the history of the examination which I was compiling. My husband's late grandmother (another Rachel) also would share her memories of hosting the aspiring candidates at their home, instead of a hostel. My mother-in-law, Prof Mary Jacob (Rtd.), shared the burden of my anxieties and the joy of every breakthrough during fieldwork and writing both the dissertation and the manuscript. Along with my father-in-law, she spread the word far and wide about my search for scholars among their colleagues.

In fact, searching for NTS scholars to participate in my study turned out to be a family pursuit with even my sister's husband, Jean, chipping in with the name of his best friend, who happened to be an NTS scholar. Ritu, my husband's brother and unparalleled computer wizard, counselled us through numerous computer woes. My sister-in-law Arin's bright smile and positive outlook lifted me from the dumps more times than I can count. Apart from all of this, both families dropped their other commitments countless times to support us in taking care of our children, so that I could work on the dissertation and subsequently the manuscript. I must also mention, my husband's uncle, Bibby Abraham, who was instrumental in arranging for the digitization of several documents pertinent to the NTS for the purpose of content analysis.

I also owe a huge debt of gratitude to several friends, who played an important role in the genesis and completion of this work. Mili and the rest of my former colleagues at the erstwhile ICICI Centre for Elementary Education (ICEE) were instrumental in spurring my interest in the issue of quality in education and the government school system. Apart from her interest in my work, I turned to Mili for sourcing literature to which I had no access in India. Sunil helped me flesh out the idea of designing a choropleth to map the NTSE data. Bobby Abrol and the whole Abrol khandaan welcomed two homeless newly weds into their family and gave us a home, love, and indispensable life lessons (including making the perfect roti, a task which I am yet to master). Sahana Ghosh helped me access rare literature from the Yale library (in between helping me pick out furniture for our home) and continues to be my go-to person for surviving in academia. Neha Gupta joined me in attacking transcription with her trademark enthusiasm and zest. Shilpa Simon sacrificed several precious holidays in helping me with transcription and also volunteered as a coder during the pilot coding phase of the content analysis. Apart from being a pilot coder during analysis, Vineeth Mathew kept me sane during many crises with an uncanny sense of when I needed her calls and prayers. The entire family at the City Fellowship Centre, led by Pastor Sam and Anitha Aunty, provided so much care, love, and prayer when I was at my lowest points.

The hidden labour of a number of women who shared various domestic tasks involved in the managing of our house at different points in time over the past 12 years must also be mentioned because their work helped me find time to work on the research. Much thanks to Asha Didi, Rakhi, Julie, Meena, Sarita, Neelam, and Susana. I owe so much to Nirmala and Mamtaji, the world's best nannies, who helped us care for our children at different points in this process of research.

I have saved the person who made it possible for this work to see the light of day to the very end, my husband, Riby. My admission to the PhD program and our wedding happened almost simultaneously. But this work would not have seen the light of day without his love, care, and 'exhortations'. I can't say I miss the years when he would work his research in the morning and care for our daughter, Joy, at night, while I clocked the night shift and cared for her during the day. But they make for several great stories. Our son, Joseph, was born towards the end of our research. Riby's pragmatic support included shouldering so many domestic responsibilities and technical troubleshooting in my countless computer woes. Our endless discussions on writing goals for the day at breakfast have become so common that it is not rare for my daughter to chide, 'Amma, aren't you supposed to be writing your book?' For the joy that this journey has been, thank you!

# Abbreviations

| | |
|---|---|
| ECR | Education Commission Reports |
| DST | Department of Science and Technology |
| GoI | Government of India |
| INSPIRE | Innovation in Science Pursuit for Inspired Research |
| IQ | Intelligence Quotient |
| IIT | Indian Institute of Technology |
| JBNSTS | Jagdis Bose National Science Talent Search |
| KVPY | Kishore Vaigyanik Protsahan Yojana |
| MAT | Mental Ability Test |
| MCQ | Multiple Choice Questions |
| MHRD | Ministry of Human Resource Development |
| MoE | Ministry of Education |
| NCERT | National Council of Education Research and Training |
| NCF | National Curriculum Framework |
| NKCR | Reports of the National Knowledge Commission |
| NMMSS | National Merit cum Means Scholarship Scheme |
| NPE | National Policy on Education |
| NSTSE/NSTS | National Science Talent Search Examination |
| NTSE/ NTS | National Talent Search Examination |
| SAT | Science Aptitude Test |
| SC | Scheduled Castes |
| ST | Scheduled Tribes |
| UNESCO | United Nations Educational, Scientific and Cultural Organization |

# 1 The National Talent Search Examination

## A History of Evolving Aspirations

> The news about my selection... mattered a great deal to me, partly because of a very silly reason. I thought I had gotten my father's approval for the first time in my life. It mattered to my father and that's why it mattered to me.

It was a cosy Delhi afternoon of October 2014, and even 37 years after he was selected as a scholar through the National Talent Search Examination (NTSE), Dev still recalled the day with sharp emotional clarity. Now a politician (whose name has been changed here for privacy), he had sandwiched an hour between meetings with members of his political party to share his recollections about his experience with me.

> .... We were a strange family in the sense that... talking about anyone's achievement was taboo. That's why when the NTSE results came, my father was pleased. He didn't say anything. But I somehow surmised that he was pleased. I thought that I had done something in my life that my father approves of.

His reminiscence smoothly shifted from an emotional to a pragmatic register as he dwelt on what the scholarship meant to him in 1977.

> It also mattered for economic reasons. It was a lot of money. In those days, it was 250 rupees a year, a month. I got three thousand rupees a year. In the first year, I was studying in my own town, which was pretty cheap. Three thousand rupees coming to me was the equivalent of something like ninety thousand rupees coming to a student today. And I was happy that I could contribute to my family's situation. Both my sisters were studying.

1977, the year which witnessed the end of the 21-month 'Emergency' declared by Prime Minister Indira Gandhi, also saw a minor development that barely made the footnotes of the history of education in India. The National Science Talent Search Examination (NSTSE), which was inaugurated with fanfare in 1964 as India's first nationwide search for talented

DOI: 10.4324/9781003344902-1

school students, quietly morphed into the NTSE. The original scheme had been the brainchild of the defence scientist, educationist, and institution builder, D. S. Kothari. Five years before he would go on to helm the eponymous Education Commission, he played an important role at the National Council of Education Research and Training (NCERT), which had just been established in 1961. Under his guidance as chair of Programme Advisory Committee for Science Education and that of his student, R. N. Rai, who headed the Department of Science Education, the institution piloted a science talent search in Delhi in 1963. It was expanded nationwide in the next year. M. C. Chagla, the jurist-diplomat who was the Minister of Education at that time, described the NSTSE as picking 'talented boys and girls who show an aptitude for science and [giving] them the finest scientific education possible' (NCERT, 1964, p. 5). The talent search's association with several prominent scientists of the period, its unique examination model, a robust financial grant for the period from undergraduate to doctoral studies in the basic sciences, and well-designed summer schools brought it much attention and prestige very quickly. But despite the early institutional enthusiasm, a decade later, it was impossible to ignore that a large number of students who were selected under the scheme did not take up the scholarship. One reason was the restriction that selected students could only take up further studies in the basic sciences. And so in 1977, the scheme was re-conceptualized as a general talent search open to those who also wanted to pursue the social sciences, medicine, or engineering. Coupled with a deliberate attempt by NCERT to publicize and to increase participation of students, this change revived the flagging interest in the program. The participation almost tripled. In 1976, only 8798 students wrote the NSTSE. In 1977, 28,955 students participated in the examination, including Dev from the small town of Sri Ganganagar in Rajasthan.

The prestige built around the persona of the selected scholars over the previous 13 years of the NSTSE leaked into the new NTSE as well. In Dev's case, the association opened the gates of Jawaharlal Nehru University (JNU) in Delhi.

> When I came to JNU, I was from a small town and this was the first time that I had encountered a metro. All the people who came [here] were from [elite] colleges [like] St. Stephen's College. …. In the School of Political Science, they were looking for a certain polish and that sort of irked me. It still does. So, I did not do very well in the interview. Just because of the way they were using language, I misunderstood them. Later my professor told me that he told them that even if the boy answered the question wrong, he is an NTS scholar. So, there is something here. Let's take him.

Dev's recollection and representation of his professor's words in many ways is a succinct expression of how commonsense and (interestingly) academic understandings of 'talent' operate, i.e. when the word is used, the speaker believes that he is not just describing the person but also predicting or explaining the person's performance (Howe et al., 1998). Talent is used in the sense of 'something innate, which makes it possible for the individual to excel' (Adamsen, 2016). The utterance *'there's something here'* is represented as being tied to a certain confidence in the National Talent Search and implicitly, by extension, to its modes of identifying the talented. This confidence in the label of 'talent' and the processes that certify its possessors in many ways continue to breathe life into the NTSE.

## Contextualizing the Talent Search

For nation-states that were gaining independence or identity in the twentieth century, the possibility of being able to identify 'talent' in children and nurture them was seductive. This was incorporated as a long-term strategy of the nation-building process in the case of several countries. For example, Israel, which became a nation-state in 1948, placed considerable emphasis on the identification of valuable human resources. The creation of a special programme of 'gifted education' (especially for disadvantaged adolescents of high intelligence identified through IQ testing) was considered crucial for Israel's national development. Another example is the former Soviet Union. From the late 1950s onwards, the USSR, faced by the imperatives of developing its space programmes and military capabilities, also developed a set of 'accelerated schools' and 'boundary schools' for the training of specially identified talented children to build a highly qualified professional cadre (Zhilin, 2010).

The independent Indian state's aspirations in this direction (as in most things) bore the imprint of its colonial past. In this case, the powerful idea of 'merit' had percolated into the collective consciousness as a possible basis for social organization through the technology and institutions of examinations in the nineteenth century. The colonial reforms undertaken by the British in education from 1854 onwards substantially altered traditional notions of 'pupil ability' and 'worth', which, for example, were tied to a familiarity with the religious canon, their interpretation according to accepted heuristics, and the assessment of the student's character (Naik, 1979). The newly instituted examination system played an important role in linking education with the competition for the achievement of status and power, such as that embodied by the civil service. Academic 'talent' and 'merit' increasingly began to be equated with examination success, where the individuals were, more often than not, tested for their mastery over the prescribed textbook content, rather than subject competence (Kumar, 1988). Since the examination was considered an objective and neutral method through which the performance of individuals could be quantified and hierarchically arranged, it provided a means to select individuals for further academic and professional opportunities based on a certain cut-off point. Those selected were considered 'meritorious'. 'Merit' functioned as a 'claim' or 'moral entitlement' based on individual performance 'in the sense of an expectation or demand addressed to the rest of the world' (Deshpande, 2006, p. 2442). The introduction of the policy of reservations for socially and economically disadvantaged groups in the selection process from the last decade of the nineteenth century further consolidated this claim. The meritorious were constructed as those who had demonstrated their legitimate and innate worth, as opposed to those who availed of reserved seats by virtue of their socio-economic background.

At the same time, one cannot also underestimate the influence of the mental measurement movement, which dominated the conceptual and research paradigms of the social and behavioural sciences, including education, at the cusp of the nineteenth and twentieth centuries (Rudnikski, 2000). The ideas of intelligence and the development of diagnostics to predict school performance, learning difficulties, and mental retardation, which originated in France and Britain, were transferred and developed in other European countries and other continents. For example, in the United States, the possibilities of scientifically measuring and tracking intelligence were seen to be applicable not only in the education context but also in the testing of Army recruits in the First World War and the eugenics movement (Carson, 2007). It was only by the first half of the twentieth century that psychology as a discipline began to gain a foothold in the Indian context, beginning with the establishment of the Department of Experimental Psychology in the University of Calcutta under the leadership of Narendra Nath Sengupta. However, it was the work of M. V. Gopalaswamy (who was trained at London University under Charles Spearman and who headed the second psychology department of the country at the University of Mysore in 1924) that popularized ideas of mental testing in India (Dalal, 2011). Indian adaptations of Western intelligence tests were developed, along with further research on applied psychology and education. The body of work that resulted fostered a notion of academic ability and intelligence that was believed to be inherent, culture neutral, and amenable to quantification. It tended to prioritize demonstrable logical-mathematical intelligence, speed in cognitive processing, and certain forms of linguistic competence which were valued by modern technological societies (Baral & Das, 2004) .

Post-independence, India envisaged education as an effective means for equalizing opportunities and reducing disparities. Therefore, the aim was to enlarge the coverage and quality of education for all, irrespective of caste, creed, sex, or economic strata. The problem however was the absence of a national vision for education. Indeed, even before 1947, there were several attempts to define a national vision for education. A significant attempt in this direction had been the resolution passed at the 1906 Calcutta session of the Indian National Congress. It asserted that the time had arrived,

> ... for people all over the country earnestly to take up the question of national education for boys and girls and organize a system of education, literary, scientific and technical, suited to the requirements of the country on national lines, under national control and directed towards the realization of the national destiny.
>
> (Naik, 1982, p. 3)

This zeal did not find a conversion to a holistic national perspective even after the first decade of independence. Education was a state subject till 1976 and the post-War Plan for Educational Development in India (prepared by the Central Advisory Board of Education (CABE)[1] under the chairmanship of Jogendra Singh in 1944) guided state initiatives in education. The first two education commissions, the University Commission of 1949 and the Secondary Education Commission of 1953 (the Mudaliar Commission), also examined only some parts of the education system. Therefore, when the NCERT was set up in 1961, there was indeed a sense that the 'national' was a perspective that had to be created and an objective that had to be accomplished. This is evident in its mandate to standardize curriculum across the country, develop textbooks, conduct teacher training, and promote research.

Unsurprisingly, the NCERT bore the imprint of both national and global agendas of the period. For example, the faculty and academics associated with the institution and its projects in the early years also brought their nationalist fervour and convictions to these endeavours because many of them were inspired by the legacy of the freedom struggle (Thapar, 2009). The faculty members were also very aware of their role as 'experts'. This awareness also owed to a predominant trend in professional socialization during the 1960s, where perspectives on desirable social change were shaped by Western 'experts'. The United States was particularly successful (as opposed to the USSR) in carving 'social territories of intellectual influence' in the Indian context, primarily because of the advantages of English education (Kumar, 1986, p. 2367). In any case, NCERT was greatly indebted to foreign 'aid' and 'advice' in the shaping of its programmes during the early years. Nine research studies in major areas of educational theory and practice were undertaken by NCERT faculty between 1961 and 1963 through the joint sponsorship of the US Office of Health, Education and Welfare and the Government of India under the USAID programme. The importance of these studies lay in the fact that they were supposed to highlight future possibilities of action for the NCERT. This effort also included the assistance of two expert teams from Ohio and Columbia in conducting the studies. Apart from this, the technical assistance from the United Nations Educational Scientific and Cultural Organization (UNESCO) and its team (7 Soviet and 2 American scientists) led to the initial formation of the NCERT's policy on science education (Kumar, 1986).

The talent search was one of the first initiatives to be spearheaded by the NCERT. The aims and the objectives of the talent search programme were defined during the initial conception of the programme as the NSTS Examination. The twofold purpose of the scheme was defined as being '*to locate the scientific talents*' and '*to provide facilities to these talented individuals so that they can blossom into creative scientists*' (Saxena, 1964, p. 2). The identification of talented students was associated with a national agenda – the need for a long-term vision and plan to develop the country's scientific competencies and staff its research institutions. The very first report put it across like this:

> With the progressive democratisation of educational opportunities and with the industrial and technological advancements achieved in the sphere of National Reconstruction, it has become imperative that the scientific potential of the country should be geared up by suitably building up a band of future research workers in basic sciences.
>
> (Saxena, 1964, p. 1)

The phrase 'national reconstruction' was one that originally found much currency among the leaders of the nationalist movement in the 1930s. The decade of the 1930s had faced a confluence of factors that demanded a re-examination of how the economy and polity should be organized. These included the cumulative experience of the economic depression during that decade, the introduction of provincial autonomy in 1935, and the growing anticipation of Indian Independence. 'National reconstruction' was a means of articulating concerns related to a 'future, possible India which could be ordered and connected with ideas of regeneration and progress' (Zechariah, 2001, p. 3690). In this light, the decision of the state to invest in 'talented students' as part of 'national reconstruction' was a significant one, especially because the channelling of resources to education was politically and economically fraught during the turbulent 1960s. The loss of the war with China, the symbolic loss of prestige in the international community, the material consequences of a massive economic slowdown, tremendous food scarcity due to the failure of the monsoon, and internal political crises created a perception of 'brokenness', which needed to be addressed (Kumar, 1986). 'National reconstruction' was a distillation of urgent aspirations of national progress.

Apart from 'national reconstruction', a second pressing reason was national anxiety about the phenomenon of 'brain drain'. The Science Talent Search report of 1967 observes: that

> this Scheme was started in response to the traditional question of brain drain. The current impact of science and technology on the daily life and on the process of National Reconstruction has made it almost imperative that we should identify a team of bright scientific and technical personnel to meet this perennial challenge
>
> (Saxena, 1967, p. 1)

The fact that 'brain drain' is qualified as a 'traditional question' in the NSTS report points to how entrenched this concern had become by the late 1960s. While race-based restrictions governed the migration of individuals to countries like the United States, Canada, Britain, and Australia in the 1930s and 1940s, specific skill shortages in these countries led to the removal of these barriers in these countries in the 1960s (Oommen, 1989).

This move was concomitant with a growing disillusionment among professionals among the middle classes regarding opportunities for employment in 'high paying and commensurately prestigious' institutions. Since the government had placed 'more emphasis on a few key institutions to provide the impetus for research and high-quality education', there were not enough places in private and public sector industries to accommodate scientists (Oommen, 1989). As I mentioned earlier, a decade later, the very nature of the talent search would have to confront the gap between the state's requirements and individual aspirations in the high attrition rates of scholars from the NSTS. From 1976 onwards, the scheme, renamed as the NTSE, was extended to cover the social sciences and professional courses like medicine and engineering.

It is important to remember that the NTSE is not the only scheme for the identification and nurture of 'talented students' under the aegis of the NCERT. The 'Chacha Nehru Scholarship for Artistic and Innovative Excellence' has been conducted by the NCERT in association with the Bal Bhawan since 2005. A hundred students are rewarded with a scholarship for 'creative performance, creative arts, creative scientific innovation, and creative writing' (NCERT, 2006, p. 77). Apart from schemes that fall under the purview of the Ministry of Human Resource Development (MHRD), other ministries also have large financial outlays to identify and nurture talented students in different areas. Two flagship programmes of the Department of Science and Technology (DST), for instance, are the prestigious Kishore Vaigynik Protsahan Yojana (KVPY), which has been conducted since 1999, and Innovation in Science Pursuit for Inspired Research (INSPIRE), which was launched in 2008. National Olympiads in science and mathematics are also held for school students by the Homi Bhabha Center for Science Education (HBCSE), under the Tata Institute of Fundamental Research (TIFR). The Ministry of Youth Affairs and Sports conducts the National Sports Talent Search Scheme (NSTSS) for the identification of talented students in sport. Similarly, the Ministry of Culture and the Centre for Cultural Resources and Training (CCRT) have been organizing the Cultural Talent Search Scholarship Scheme (CTSSS) for children of the age group 10 to 14 to develop their talents in various artistic fields. In addition to the 'talent search- scholarship' model, there are also school-based interventions for supporting the 'talented'. In 1986, the National Policy on Education recommended the creation of special residential schools for academically talented students from rural areas in every district of the country, i.e. the Jawahar Navodaya Vidyalaya Scheme. Various state governments also have their respective schemes. For example, the Delhi Government has established a chain of special Pratibha Vikas Vidyalayas for 'talented' students who are selected from government schools. Apart from government initiatives, there are also several private institutions which nurture the talents and gifts of 'exceptional' students, such as the Jnana Prabodhini Prashala of Pune.

## Research on the Talent Search

The National Talent Search is now more than 50 years old. However, one of the astonishing features has been the lack of research attention that it has attracted outside of the NCERT. Nearly all the available (and few) studies on the NTSE in the public domain have been conducted by NCERT faculty or through financial assistance from the council. What is common to these studies on talent search is that they work within a paradigm that acknowledges the existence of a quality called 'talent', the necessity for its identification, and recommendations for the improvement of the process in exams like the NTSE.

The first type of research on talent search is those that seek to explore the relationship between the design of the examination and the achievement of its ends, i.e. the identification of the talented. In this context, M. K. Raina's *The Talented Scholars (1991)* is the most holistic study of the NSTS which has been done so far. It examines the objectives and impact of the scheme using a longitudinal study of 130 NSTS scholars from the batches of 1964 and 1965. The study builds a psychological profile of the selected scholars. Raina's startling conclusion was that his subjects *did not* show behaviour patterns that were significantly different from those who were not labelled as talented. With a few exceptions, he assesses his subjects as not having shown very high levels of mature achievements and accomplishments. He argued that their personality assessments did not match those of the 'truly creative', because they display an authoritarian value system, characterized by a feeling of inferiority, powerlessness, and individual insignificance, with a tendency to depend on powers outside himself or herself. The sources of the authoritarian syndrome are traced by Raina to the family and the socialization patterns of Indian society. He does not mince words in using these findings to argue that the programme could not identify the 'truly talented' because it did not conceptualize talent as a psychological entity that was to be identified and nurtured. He argues that this problem was further compounded because of the programme's 'one dimensional and uniform view of the human mind'. Therefore, the non-intellectual determiners of talent such as 'task commitment, including perseverance, endurance, hard work, dedicated practice, self-confidence and a belief in one's ability to carry out important work' (Raina, 1991) found no place in the identification of the talented. In this rather scathing assessment of the program, the real stinger is his conclusion that these individuals could have achieved what they did even without the talent search award. Recommending an overhaul of the programme in its current form, Raina argues for the need to recognize diverse signs of excellence by using multiple criteria for the selection process as well as using multiple entry points into the programme to reduce exclusions and broaden participation.

In another study conducted along with Srivastava, Raina extends this argument to various other talent search programmes in the Indian context, i.e. the lack of an adequate policy perspective on what constitutes talent impairs 'India's search for excellence' (Raina & Srivastava, 2000). Vani Geddam makes the same argument in her study of the Navodaya Vidyalaya Scheme (2003). She also affirms the existence of a psychological entity called 'talent', which must be understood in its complexity before 'truly deserving students' can be identified through such programmes. Like Raina, the aim of her study is also to promote a wider understanding of 'talent' and 'excellence' beyond mere academic ability. This requires the adoption of multiple pathways of identifying the 'talented' than just testing verbal and logical intelligence (Geddam, 2003).

In contrast, Yagnamurthy (2010) assesses the NTSE by returning to the original perspective which governed the creation of this programme, i.e. the need for excellence in various disciplines and the role of the state in channelizing the talented into suitable streams. In 2007, he studied 51 students who were participants in a five-day nurturance programme conducted for awardees from Eastern India (West Bengal, Orissa, Chhattisgarh, Jharkhand, and Bihar). He used their mean score of 94.02% in their Class X examination to demonstrate that students selected through the talent search were also highly successful in the Board Examinations of their respective states. Twenty-one of these awardees desired to pursue a career in engineering, followed by nine who expressed an interest in Science Research and seven who wanted to go in for medicine. None of them considered pursuing a career in the social sciences or humanities. He also notes that between 1985 and 2004, most of the NTSS awardees dropped out of the scheme after the undergraduate level instead of proceeding to their post-graduate degrees. His analysis is that the availability of employment opportunities as well as remuneration for engineering graduates post-liberalization and privatization in the 1990s has led to a decline in the number of students pursuing the basic sciences, arts, and commerce. He argues that the play of market forces in attracting talented youngsters to certain disciplines like medicine and engineering, as opposed to others, must be checked by governmental attempts to create a favourable educational and professional ethos that draws them to the basic sciences, social sciences, and humanities. If the NTSE were to play this role, it would have valuable significance (Yagnamurthy, 2010).

A second type of research focusing on studies on talent search has been oriented around the social and economic composition of the beneficiaries. Raina's review of the National Talent Search Scheme indicates that talent search tends to favour urban, male candidates from socially and economically privileged groups, studying in schools with good infrastructure and English as a medium of instruction. These children tend to belong to families with relatively high educational and income levels, as reflected in the backgrounds of their fathers who would mostly be engaged in technical, professional, or administrative jobs (Raina, 1984, 1989). The notable point is that this phenomenon does not just seem to be limited to the NTSE. A similar result was also observed in the case of the Kishore Vaigyanik Protsahan Yojana, which is conducted by the DST, and seeks to identify students with a special aptitude for science. P. S. Kumar and Dipankar Chatterjee's decadal review (2009) of the scheme raised the point that only around 20% of successful candidates are women.

An appraisal of the NTSE-2001 done by Jain (2002) continued to reflect this trend, including the huge skew in terms of representation between boys and girls. The responses of 3888 candidates in the 2001 NTSE's Mental Ability Test (MAT) and SAT (Subject Aptitude Test) were examined to assess the influence of gender, area (rural/urban residence and school), and caste categories on the total scores of MAT and SAT, the scores of subtests of MAT and SAT, and interview marks on the 3888 candidates. A comparison of their mean scores revealed that MAT was found to be easier than SAT. With the SAT, it was found that mathematics had the least mean score (10.1) and that the social science subjects (history, geography, and civics) had the highest (24.8) (Jain, 2001). Jain demonstrates that General Category (GC) students tended to perform better than Scheduled Caste (SC) and Scheduled Tribe (ST) candidates in both SAT and MAT, that boys outperformed girls in both, and that urban candidates fared better than rural candidates. The study highlights these differences but does not investigate the cause of these divergences or contextualize it through a socio-economic analysis. Rather, the greater focus was on demonstrating the merits of the testing instruments. For instance, the better performance of the awardees over those who were called for the interview is used to demonstrate that the two-tier system (written exam followed by the interview) does discriminate between the two groups. The reliability of the MAT and SAT was also demonstrated to be statistically high (Table 1.1).

In this context, Pandey (2004) offers more insight into the causes which perpetuate such patterns of exclusion in his examination of the causes for the poor participation and achievement of girls from Uttar Pradesh in the NTSE. Conducted under the aegis of RIE, Ajmer, the study covered 56 schools of 14 districts chosen from 8 divisions of the state (Meerut, Moradabad, Bareilly, Faizabad, Lucknow, Kanpur, Chitrakoot, and Jhansi). In all, 815 girls were selected (344, who appeared for the NTSE, and 471, who did not) for the study. A major finding was that 48% of the total students who were part of this study did not know about the scheme. Pandey

*Table 1.1* Demography of NTSE-2001 Winners

| The NTSE (2001) | Boys | Girls | GC | SC-ST | Urban res. | Rural res. |
| --- | --- | --- | --- | --- | --- | --- |
| Appeared for written exam (Total: 3888) | 3087 (79.4%) | 801 (20.6%) | 2956 (76%) | 932 (24%) | 3241 (83.4%) | 603 (16.6%) |
| Called for the interview (Total: 1533) | 1236 (80.6%) | 294 (19.4%) | 1181 (77%) | 352 (23%) | 1327 (86.5%) | 198 (13.5%) |

(Note: Data compiled from Jain (2002))

argues that the poor performance of girls in the NTSE needs to be understood in terms of the poor dissemination of information about the examination leading to very low participation and inadequate preparation. At the level of the district administration, the officials admitted that the burden of other administrative responsibilities, the lack of finances as well as staff, prevented them from conveying the notice of the examination and the availability of the application form. This, in turn, led to poor turnout for the test. Pandey notes that the apathy regarding the test also percolates to the school level, where principals tend to place the onus of finding out the dates, procuring the forms, and preparing the students for the exam on the teachers. The burden of other duties, coupled with a lack of support from the principals, leads to teachers not prioritizing this exam. The majority of the parents of both students who did not clear the exam as well as those who did not appear for it had very little information about the test. Therefore, it is not surprising that the participation of rural and government school children in the exam was less than 1% of the total eligible enrolled students. Those girls, who do participate, tend to be from the district headquarters and appear through their own effort. Concerning the test itself, the compulsory section on mathematics in the SAT was a big challenge to the girls, as most were from girls' schools that offer home science instead of mathematics at the Secondary level. The ability to prepare for this examination is also limited by poor libraries in rural schools and the lack of literature that offers an understanding of the model of the questions (Pandey, 2004).

Pandey's study seems to echo the concerns of an earlier article written by Nambissan and Batra (1989) concerning the perspective of the State in implementing 'talent' development programmes. In their analysis of the Navodaya Vidyalaya Scheme, which set up special schools in every district for talented rural students, they argue that the manner in which the programme is conceived, the stage at which it is to be implemented, and how the children are selected ensures that the interests of the more privileged strata in rural India are served. The emphasis on fostering 'excellence' is seen as a smokescreen to rationalize the shift in the priorities and resource allocation of the State away from primary education. Secondly, through an analysis of the testing instruments used to select the students to the Navodayas, they demonstrate that it is not innate cognitive abilities or natural talent which are likely to be reflected in the test performance but largely the previous learning experiences that children have been exposed to (Nambissan & Batra, 1989).

The common characteristic of the studies, which have been described above, is that they approach the study of a talent search programme from the perspective of 'talent' – its constitution, its recognition, its development, etc. – or from the perspective of 'social equity'. As is very evident, this area is indeed fertile for exploration in the Indian context, considering that very few studies are devoted to studying how talent is understood in government programmes and that the NTSE itself has hardly received any research attention outside the NCERT. As many of the studies reviewed above noted, there is a great conceptual vacuum at the level of policy with respect to talent. They also rightly draw our attention to the inadequacy of solely focusing on the question of 'talent' because it leads us to miss how powerful social discourses and practices (gender, caste, nationalism, etc.) influence the identification and nurture of the ability of students.

Yet, as the history of the changes in the test as well as the available studies of the NTSE demonstrate, it is evident that there has been considerable anxiety about whether the test has been fulfilling its mandate of identifying the truly 'talented'. But is it not interesting that this anxiety has not been considered sufficient to abandon this project of identifying the 'talented' for a larger social good? For instance, while the search for scientific aptitude is justified quite vigorously in the official documents in the 1960s, the absence of a theoretical grounding or reflection on the implications on the shift from a search for 'scientific aptitude' to generalized 'mental ability' or 'scholastic aptitude' in the 1970s is striking. This does not seem addressed in any of the internal studies that have been conducted post-1976. Despite the trenchant critiques that have been levelled against it, the idea of a talent search continues to be accepted as something which is desirable for the social good within the discourse of the NCERT and that seems to be sufficient justification for its existence. Therefore, the existence of the test, as well as the continued investment in it, begs the question of what function it performs within the State imagination. So what is this special quality that seems to attach to the very idea of 'talent'?

The word 'talent' is also not rigorously defined according to the conventions of psychology in the NTSE literature. For example, the concept of 'talent' in the NTSE seems to be treated differently from that in other schemes under the NCERT itself such as the 'Chacha Nehru Scholarship for Artistic and Innovative Excellence'. The emphasis on creativity in the latter scheme contrasts with the total absence of the word in relation to the current discourse of 'talent' with the NTSE. A perusal of official documents related to the NTSE (on which I will elaborate in the next two sections) allows us to see how the word 'talent' implicitly circulated in association with abilities, aptitudes, mental endowments, worth, potential, excellence, desire, and so on.

Despite the ambiguity associated with the word, it seems to have the felicitous quality of being able to generate consensus. For instance, J. P. Naik, who served as the member secretary of the seminal National Education Commission (the Kothari Commission) of 1964 to 1966, noted that amidst the highly contentious reception of the final report, the identification and development of talent was one of the few proposals which attracted wide attention but which did not cause controversies. '*There was agreement on the fact that 'top bracket students should be the wards of the state which would assume all responsibility for their education*' (Naik, 1982, p. 52).

Another lacuna in existing studies of the talent search is the absence of an exploration of how 'talent' itself gains meaning when it is attached to the lives of individuals and internalized by them as a category that allows them to make sense of experiences and interactions. For instance, the NTS scholars were those who were selected when they were in school and before their potential had been converted into socially meaningful contributions. It is worth exploring how their experiences reveal a different facet of socialization towards the State. Of related significance is the perception of the NTSE in the public imagination (its visibility as a locus of desire in some sections of society leading to investment in coaching and preparation for success in the exam and its invisibility in other social segments as traced by Pandey above). These were some questions that I grappled with as I embarked on studying the NTSE.

## An Alternative Approach to Talent

When I began this research, my interest was in trying to understand how 'talent' is a label that can be won through success in an exam like the NTSE and how its prestige makes it a locus of tremendous social desire. I wanted to explore how selected scholars represented the impact of being recognized as 'national talent' on their personal and academic trajectories. What kind of traffic existed between a certain symbolic certification and the material reality of their lives?

However, the paucity of studies on the NTSE programme necessitated a rethink of this initial direction. I quickly realized with dismay the NCERT had no documents that consolidated a history of this programme, how it evolved, and why it followed certain directions but not others. Additionally, the idea of beginning the study through exploring the perspective of a sample of former winners hit a roadblock on the discovery that there was no updated master list of all the winners of the programme from 1964 onwards. It was evident that considerable groundwork needed to be done to establish the history of the examination and the profile of its winners. As I poured over annual reports, disparate official documents, policy texts, and interview narratives to construct an archive about the examination, my focus turned to the language which is used to talk about notions of talent and represent knowledge about it, i.e. the discourse associated with it.

To explore the idea of 'talent' in this way, I had to distinguish it from how it is usually explored and measured in relation to cognition, learning, creativity, personality, social and emotional adjustment, gender, etc (Philip, 2016). While these are important entry points, to look at 'talent' as a discourse was to consider language as social practice, a way of conceiving and analysing the internal and dialectical relationship which exists between language and society (Fairclough, 1989). This is what makes it possible to build a topic in some ways while delimiting other possibilities. All discourse is more than language texts, spoken or written. Equally significant are the processes of the production and interpretation of language texts. While individuals indeed use cognitive resources to interpret a text, these resources also have social origins, i.e. 'they are socially generated and their nature is dependent on the social relations and struggles out of which they were generated- as well as being socially transmitted and, in our society, une-qually distributed' (Fairclough, 1989, p. 24). Therefore, discourse practices are inextricable from social practices, i.e. social conditions of production and interpretation. These conditions may also be seen as related to specific levels of social organization. For example, it may be the level of the social situation or the immediate social environment in which the discourse occurs. Or it may be the level of the social institution in which the discourse is embedded. It may even be the level of society as a whole.

As I pieced together the history of the NTSE, I began to notice that the representation of 'talent' and the program's conception and development in the official documents were inextricably connected with that of a second discourse, i.e. that of the 'nation' (and to be specific, 'nation-building'). At this point, it is important to remember that the 'nation' is also a cultural construct sustained by discourse. The basic premise which underlies this assumption is that nations are imagined into existence through concrete social mechanisms. These mechanisms produce a specific kind of conscious-ness that allows individuals to identify themselves with countless, faceless others in an experience of a 'we' feeling (Anderson, 1991). This 'national' identity has to be continuously built, sustained, or negotiated in the face of other discourses which may undermine or even threaten it (Bhabha, 1994). When the word 'national' is attached to 'talent', 'talent search', and 'talent scholar', these concepts participate in the substantiation and affirmation of the 'nation'. Therefore, the discourse of 'talent' both constructs and is 'con-structed' by the discourse of the 'nation'.

Greater familiarity with the official literature (scattered as it was) soon made it obvious that a reconstruction of the history of the talent search was a better entry point than beginning at the level of the winners of the schol-arship. A unique aspect of the NTSE lay in the fact that it is the only national-level test conducted for school students under the aegis of the Ministry of Education and the NCERT. Therefore, the discourse of talent within this examination is closely connected to the larger policy perspective of the country on education. This is important because it is at the level of policy that the 'talented student' is first constructed as a special category among students who require special support from the state (Schneider & Ingram, 2008).

Since the students were selected into the program when they were adoles-cents, I increasingly began to see that the NTSE offered an opportunity to explore the relationship between the state and the citizen. Paying attention to the construction of the idea of 'talent' in a government programme was an opportunity to understand the state's ability to create new identities which possess tremendous symbolic capital. This capital is built on the per-ceived oneness between the state and the nation. To be recognized by the state is to be recognized 'nationally', and this is a source of great prestige. Here, let us note that an individual who is a National Talent Search Scholar possesses prestige, which is different from the winner of a talent search contest which is organized by a private body (the multitude of private

Olympiads or competitive talent search programmes including reality shows like 'India's Got Talent'). While the latter may also invoke the idea of an 'all India' competition, the difference in prestige lies in the fact that the former is a discourse of the 'nation', which is bolstered by the power of the state. The label of 'national talent' and all that it represents at a given point of time is the raison d'etre for the investment in and nurture of a select number of students by the state. This entails expectations from them in terms of contributions to the Indian economy, polity, and society. Therefore policy designs do have a pedagogic role in teaching citizens how they can be expected to be treated by the state (Schneider & Ingram, 2008). Individuals do not automatically think or act 'nationally', i.e. as a 'member of the nation'. Rather they are socialized into such thought and behaviour patterns through the continuous work performed by the institutions of the modern state in creating and sustaining a national consciousness among the members of its body politic.

With this kind of framework in mind, I began to elicit narratives from a select number of winners between 1964 and 2000. As I began speaking to scholars, it was evident that the conceptions and opportunities which were created by the state's search for the 'talented' affected their self-concept and pursuit of self-fulfilment in unexpected ways. Interviews with winners of the scholarship continued to remind me throughout this journey that statist ideas of the relationship between 'talent' and the 'nation' are very different from the lived experiences of ordinary (and extraordinary) individuals.

To sum up our discussion thus far, this book uses the quiet history of the NTSE examination to undertake two related explorations. One trajectory explores the relationship between the ideas of talent and nation-building through the institution and the evolution of the NTSE in post-Independence India. This works as a scaffold within which is nested a second exploration on how personal conceptions of talent are related to institutional ones. To do so, I draw on memories and perspectives of National Talent Search Scholars between 1964 and 2000 to reflect on the similarities and differences between personal and institutional constructions of 'talent' and the 'nation'.

## Assembling an Archive

The absence of organized information about the study made data collection a challenging process. As I mentioned, I embarked on this research, cheerfully assuming the existence of an archive of information about the NTSE. Quickly disabused of this notion, data collection took the form of creating an archive of texts, which would form the basis for a reconstruction of the programme's evolution and an interpretation of the same. By texts, I include both physical texts (literature pertinent to the study) and the narratives of a sample of NTS scholars which were obtained through semi-structured interviews. The way data was collected had a critical impact on how the NTSE became an object of reflection and debate. My agency as a researcher, the kinds of materials I had access to and didn't, the sites from which the documents were collected, the choice of texts which were considered significant or otherwise, the kinds of data and narratives which were generated because of my research interest, etc. is an integral part of how an archive regarding the NTSE was formed.

As I worked on this, my broad aim was to build an archive of secondary sources on the NTSE from its inception in 1963 to 2013 to describe and analyse the evolution of the programme. I wanted to identify patterns in the national and state-wise participation of students in the NTSE and the distribution of winners to understand the profile of a 'typical' NTS scholar. I hoped to identify and interview a select sample of NTS scholars (based on year of selection, choice of undergraduate subject, location while writing the examination, current profession, and gender) and to document their experiences of the programme. As I began to get a better handle on the representation of the 'talented student' in these narratives, I sought to compare this with the kind of language used in the Indian education policy documents after independence.

The sources to which I turned for information on the NTSE were not all simultaneously available at the start of the study. But the guiding imperative behind the selection of these sources was how they helped answer various questions which emerged at different points in the study. The institutional literature published by the NCERT was the starting point. Apart from that, the rest of the sources of information were those which helped me piece together missing pieces in the narrative which was emerging from the NCERT literature. The sources, as they are listed below, are not arranged chronologically with respect to how I found them. Rather, the structure of this listing reflects how I finally pieced together different aspects of the history of the examination.

### Institutional Literature Published by the NCERT

I mined 50 annual reports of the NCERT from the time of its inception (the first one being published in 1963) to 2013 for a bare-bones but continuous documentation of the talent search. Each report had a few pages on the NTSE. The most important information which they contained was the tabulation of the winners from every state for each year. It was possible to trace the national distribution of 20,678 winners between 1964 and 2004. The reports were also the source of information regarding changes within the scheme. The earliest reports showed much evidence of wear and tear. In some years, the English versions of the reports were either lost or not stored with the rest of the report collection. In those cases, I turned to the Hindi versions of the reports for the relevant years.

Details regarding the inception of the talent search programme were obtained from five reports published by the NCERT between 1963 and 1967. They were written by K. N. Saxena of the Department of Science Education (which originally conducted the test). Only the reports of 1966 and 1967 could be traced at the NCERT. The earlier reports for the years

1963, 1964, and 1965 were found in the library of the Central Institute of Education (CIE), University of Delhi. A further search at this library unearthed a compilation of essays written by winners of the NTSE in 1963. Locating these documents in this library was a reminder of the forgotten history of the incubation of the NCERT at the CIE during the former's early years. However, poor maintenance here also has led to the destruction of several other documents from the early 1960s, which are indicative of this shared history of institutions. In addition to these, those internal studies of the NTSE undertaken by NCERT and RIE faculty (between 1963 and 2014) which could be traced were also used as sources.

### Lists of Winners

While the Annual Reports of the NCERT provided a state-wise distribution of winners, what was not evident was if the winners were clustered in certain parts of the states. This information entailed accessing details of the winners of the previous years. A significant roadblock in retracing the history of the test was the discovery of the absence of a master list of winners of the scholarship since its inception. In my first request to view the entire database of winners and their addresses on June 28, 2012, I was declined permission to view the records of names of winners, as this was 'confidential information'. I was asked to crowd-source the names of winners through social media like Facebook. As a token, the names of a few 'famous' winners were given to me. However, through the assistance of the Coordinator of the Test and a former Head of the Department of Education Measurement and Evaluation, I discovered that the names of winners after 2000 were accessible in handwritten registers and from 2008 in the form of a computerized database. I requested and was granted permission to view one of the registers of scholarship disbursement post-2000. I was first given access to the 2005 register, which contained a list of the 1000 winners selected that year. Each winner's name was followed by his address, which usually contained a reference to his or her village, town, or city (or in other cases, pincodes). Since the register was handwritten and had no digital counterpart, I manually classified the data, leading to a district-wise distribution of 4811 winners between 2001 and 2005.

### Information Secured through the Right to Information Act

The data of winners begged more questions. Did the district-wise distribution of winners from states correspond to the district-wise distribution of participants from each state? In fact, participation data was referenced in the Annual Reports of the NCERT only up to 1984. Since the test was decentralized from 1985 (with a preliminary state and a subsequent national exam), the NCERT officials had told me that the participation data was stored at the state-level liaison offices, which conduct the first state-level examination.

Finding the online application forms for the NTSE examinations between 2009 and 2013 was another unexpectedly productive avenue of enquiry. These were invaluable sources for understanding the kind of information that was being collected on the participants. However, the difficulties encountered in securing access to the NCERT records of winners shaped my decision to use the Right to Information (RTI) Act to seek the state-level participation data. Using these, I drafted a basic RTI application with nine specific queries for the data collected each year between 2009 and 2013. I located the online list of regional liaison officers for the NTSE published on the NCERT website and filed 35 RTIs to the institution to which each officer was attached, i.e. one to every state and Union Territory. I obtained responses from 19 states (excluding Punjab, Orissa, Madhya Pradesh, Gujarat, Assam, Meghalaya, Delhi, Manipur, and Mizoram) and 4 Union Territories (including Chandigarh, Pondicherry, Daman and Diu, and Dadra and Nagar Haveli). The responses were varied in how they answered my queries. Predictably, the majority of the states got back to me with responses to just one or two questions but some were very detailed and informative.

### Autobiographical Accounts by NTS Scholars

My first exposure to the worldview of the NTS scholar was rather fortuitous. While browsing through the Delhi University bookstore, I stumbled upon a book titled *The Girl's Guide to a Life in Science* (Ramaswamy et al. 2011).[2] Of the many accounts by women scientists in the anthology, two of them mentioned that being selected in the National Talent Search was momentous in their academic journeys. To understand the way these narratives fused new energy to this research, let me share an extract from one of them, a molecular biologist, Sudha Bhattacharya, who won the NTSE in 1968.

> ..... Through most of school I more or less viewed science as a subject that allowed one to get good grades, rather than a source of wonderment or creative joy. It was only in class eleven, when we were introduced to genetics and the DNA double helix, that I felt myself participating in the thrill of scientific discovery. For the first time, I found it exciting to move beyond merely accepting the formulae laid down in the textbook- to actually stop, wonder and question. Even so, I did not yet envision myself as a scientist. It was only when we attended our first National Science Talent summer school that I first saw how timid and constraining this approach was. When we met the contingent from Bombay, it became clear that they had far more exposure than our Delhi Group. They had encountered internationally renowned scientists visiting Mumbai for lecture tours and knew about thrilling discoveries in molecular biology. Suddenly it occurred to me that perhaps

there were more challenging avenues to explore than school teaching!
..... My heart was now firmly set on molecular biology- I found it
impossible to reduce the beautiful and myriad differentiation processes
in plants to mere 'life cycles' that one mugged up.

(Bhattacharya, 2011, p. 11–12)

I would encounter this kind of affective engagement demonstrated ('source
of wonderment', 'creative joy', 'thrill of scientific discovery', 'wonder',
'thrilling', 'challenging', 'heart', etc.) with the Talent Search Nurturance
Programme (its summer schools) in other accounts by winners from the first
decade of the program too. Most of them drew attention to the role played
by the nurturance program and summer schools of the NTS. Three NTSE
winners from the first decade wrote about the mentorship of the eminent
scientist and educationist, Prof B. M. Udgaonkar in a volume dedicated to
him by the HBCSE. They wrote of how he used to invite NTS winners from
undergraduate colleges in Mumbai for weekly discussions on science
problems.

I discovered yet another poignant source in a blog which was written by
Joseph Pinto, a winner of the NTSE from 1968. In an extensive piece in
which he reflected on turning 60 years old, he devoted considerable space
to the place of NTSE in the trajectory of his life. Around the same time,
I discovered the most high-profile endorsement of the talent search, i.e. on
the Nobel Prize Website. Writing about his academic journey, Venkatraman
Ramakrishnan, the winner of the Nobel Prize in Chemistry in 2009, reflected
on the role played by the NTS in his pursuing a career in science, rather
than in medicine or engineering. These narratives revived my interest in
understanding the examination's legacy from the perspective of the winners
and were useful in designing the interview questions for the scholars who
eventually participated in the study.

## Interviews of NTS Scholars (1964–2000)

In June 2012, my dream of finding and contacting a random sample of
winners enjoyed a brief resurrection. A section officer of the office of the
disbursement of the NTSE scholarship, who had averred that records of
names before 2000 were not available, contacted me saying that he had
discovered a cupboard with some old files with names of winners. However,
I was warned not to have much hope since the material in the cupboard was
in such poor shape that a bad smell was emanating from it. I observed that
the details of the winners prior to 1999 were contained in dilapidated A4
size notebooks of very poor quality. They were also not organized year-
wise, though they were kept in stacks. In some cases, the lists of winners
were handwritten and in others, typewritten lists of winners were pasted
onto their pages. Pages were brittle to the point of disintegration and were
often loose. In fact, as some of the books were lifted out for my perusal,

a few pages fell out and I witnessed another staff member stuff them into a random notebook at hand, careless of the worry that such an act distorted the records.

I had hoped to understand how the data about the winners is organized so that I could figure out how to constitute my sample. With a view to understanding the changes in the perspectives of the beneficiaries, I planned to identify winners from 1964 to 2004. 2004 was fixed so that four decades could be earmarked from the beginning as well as to obtain winners who had progressed far enough in their educational and professional trajectories so as to be able to reflect on the influence of the NTSE in their lives. As a pilot exercise, I wanted to obtain a minimum of 100 names and then work with that data.

The official who revealed these notebooks graciously offered to photocopy the pages that I needed. The names and notebooks were randomly selected. I hoped to obtain names from a broad spectrum of years and at least make sure there were a few names from each decade from 1964 to 2004. The first page of each register selected in this fashion was photocopied. I obtained 21 such lists and therefore, around 200 names. The names of the listed scholars were then sought online. Eventually, I obtained 37 possible matches out of the 200 names. Only 5 of them explicitly mentioned winning the NTSE in their resumes. Others were possible matches, traced based on other details in their profile, such as years of schooling, college and year of graduation, undergraduate course, and geographical location. Email ids were rare finds. Many were located through their online LinkedIn Profile. Additionally, out of these 37, only 10 could be confirmed as residing in India, based on the online information.

Finally, I turned to writing to convenience-based sampling by writing to NTSE scholars whom I discovered through Google searches, LinkedIn, blogs, books, etc. and snowballing others through contacts from the scholars I interviewed. Technology and online platforms such as Skype and Google Hangouts were critical in identifying, verifying, recruiting, and interviewing participants (who were so widely dispersed in terms of geography) for the study.

Out of a total of 43 individuals who were contacted, 30 responded. (At the very minimum, they acknowledged the receipt of the mail). Nineteen out these 30 eventually agreed to participate in the study (11 men and 8 women). Eleven of these 19 scholars had been contacted by me either after reading an article, blog post, or online resume entry which mentioned their winning the NTSE. The remaining 8 were obtained through references from those who were already interviewed. Ten of these individuals are recipients of the NTSE award between 1964 and 1970. Seven are recipients between 1971 and 1980. The remaining two awardees are from the 1984 and 2000 batches, respectively. Graciously, they shared their stories with me, filling in gaps and raising new questions about the historical data that I had pieced together thus far. The names of the scholars have been changed throughout this account.

*Educational Policy Documents*

The last textual source to which I turned in order to supplement the information gained from the NCERT and the narratives of the scholars were
policy documents. While this would have seemed a natural source of information, it took time for me to see the interconnections between the procedural language used by the NCERT, the personal language used by the
scholars, and the aspirational language of the policy.

A total of 23 policy documents of the Government of India (between
1947 and 2010) were examined for their discourse on talent. Of these, 11
were five year plans prepared by the erstwhile Planning Commission
(between 1951 and 2012). Apart from this, the remaining documents were
the two National Policies on Education (1966 and 1986), the Programme
of Action (1992), the paper Challenge of Education (1985), the three
National Curriculum Frameworks (1988, 2000, and 2005) and significant
education commission reports including the Secondary Education
Commission Report (1952), the three volumes of the Kothari Commission
Report (1966), the Report of the Ramamurthi Committee (1990), and the
Yashpal Committee Report (1993). The reports of the National Knowledge
Commission from 2006 to 2009 were also examined. All these documents
were hosted at the Teacher Education portal of the Ministry of Education
(formerly the Ministry of Human Resource Development) website, except
for the paper 'The Challenge of Education'. From these 23 documents, the
11 which had a greater focus on the discourse of talent were chosen for
closer analysis. These documents provided a picture of the evolution of the
discourse of 'national talent' in the larger discourse of the state. The new
National Education Policy of 2020 was not an original part of the study.
However, some reflections on this document are also included.


## 'There's something here'

Having mapped the raw material with which this study worked, now let us
revert to a bird's eye view of where these explorations will take us. I do
not begin with a definition of 'talent'. Over the preceding four sections of
this chapter, I have sought to make a case for how one may understand
'talent' as a discourse from a social and historical perspective, in contrast to
beginning with psychometric models of identification and evaluation.
Rather, this book intends to contextualize and historicize the explicit and
implicit understandings and assumptions that nest within this discourse,
aspects which are intuitively alluded to in statements like the description of
his ability that Dev attributed to his professor, i.e. 'There's something here'.
The subsequent chapters will elaborate on the theme by exploring some
aspects of the history of the institutionalization of 'national talent' in
post-Independence India.

To understand the articulation of these three words, 'talent', 'nation', and
'science', we need a history of ideas such as attempted by the cultural

theorist Raymond Williams in his work on 'keywords', where he traced 'significant binding words' in certain activities... and 'significant, indicative words' in certain forms of thought (Poulson, 1996). As a step in this direction, the second chapter traces the linguistic transformations in the discourse of talent between the fourteenth and twentieth centuries in the Euro-American contexts and their influence on the aspirations of nationalist leaders for the independent Indian nation. In doing so, a conceptual framework is created for understanding the relationship between talent and nation-building that shaped the genesis and subsequent evolution of NTSE.

The third chapter continues this project of creating a context for the evolution of the talent search through a close examination of Indian education policy after 1947. A close analysis of the selected policy texts reveals the sheer marginality of the discourse of 'talent', and references to the word comprise less than 0.01% of their word count. Yet, among them, the Report of the Education Commission 1964–1966 (1966) stands out for having 43% of the total references to the word 'talent' among these documents, followed by the National Knowledge Commission Report to the Nation 2006–2009 (2009) with 21%. Separated by a span of 40 years, a close survey of how 'talent' is represented in both reports offers crucial insights into how this concept is constructed in policy and how it has evolved in post-Independence India. A morphological content analysis of the language associated with 'talent' in both documents (i.e. the use of verbs, adjectives, collective nouns, and synonyms) exposes some of the active meanings and values that are embodied in this discourse. This enables us to reflect on some continuities and breaks within the state's imagination of its relationship with the talented citizen in the project of nation-building.

We move next to the interpretation of policy at the institutional level by recovering some clues to how NCERT imagined the 'talented student' in the genesis of the science talent search and in its subsequent transformation to a general talent search. Inspired by the image and involvement of 'nationalist-scientists', the genesis of the science talent search is described in detail in the fourth chapter. The programme experimented with novel technologies of testing that significantly affected the form, content, and vernacular of competitive examinations in India. This included the use of 'multiple choice questions', the categorization of some questions as 'thought-type', the assessment of 'extra-curricular' knowledge as a proxy of ability and interest, etc. At the same time, the marginality of the discourse of talent in Indian education policy necessitated considerable work in justifying an intervention for the talented. This is evident in some of the arguments used to build a case for a special 'nurturance program' and included the critical age of intervention, the construction of the general student population as a 'problem', and the expressions of a lack of confidence in teachers and higher education institutions. A critical analysis of the short history of the NSTS enables us to reflect on the institutional work done in constructing the category of 'science talent' and creating opportunities for this special cohort of students under the aegis of nation-building.

This evolution of this institutional imagination of nation-building is further explored in the face of the fact that the design of the scheme fell short of the expectations of both policy-makers and the aspirations of those who were selected. Chapter 5 traces how the redesign of the examination shifted its emphasis from the identification of talented students for their nurturance to a conception of the act of identification as an end in itself. This was characterized by attempts to make the exam more formally inclusive, with an implicit institutional reconfiguration of the idea of nation-building to one based on expanding the social geography of the talent search at the formal level. The complete lack of articulation of the envisioned impact on the selected scholars as well as the steady decline in their academic nurture speaks to the shift away from conceptualizing nation-building through the contributions of the talented. To substantiate this claim, this chapter analyses patterns of the profile of the NTS scholar under both avatars of the talent search using national, state-wide, district-wise, gendered, and caste-based data of winners and participants, as well as the academic choices of scholars.

The sixth and seventh chapters introduce a different perspective on the same history of the scheme – the composition of the examination, the elements that were part of the program in the initial science talent search decade, the ramifications of the changed scheme, and the nature of its legacy – from the perspective of a small sample of winners (11 men and 8 women who participated in the program between 1964 and 2000). Their accounts based on their recollections and reflections provide a counterpoint to the state's imagination of the talented. While the sixth chapter specifically locates their participation in their program as a means to reflect on the occupational conceptions of individuals during late adolescence. In their narratives, these occupational conceptions are also tied to an emergent understanding of 'aptitude' as a readiness that is shaped and reinforced by their experiences in their families and at school. But how do these individuals interpret the concept of 'talent' and its relationship to nation-building? The seventh chapter puts together their reflections on the nature of talent using their responses to different aspects of the program's evolution (the role of science in nation-building, equity and affirmative action in educational selection, the relationship between the state and the student, etc.). The relationship between self-knowledge, privilege, and agency which emerges in these reflections is engaged with in-depth.

The concluding chapter pulls together the various threads that were drawn from policy texts, institutional records, and narratives of winners: the question of academic talent; the state's role in its constitution, definition, identification, and nurture as an aspect of 'nation-building'; and the rights, expectations from, and aspirations of the student citizen. By exploring the roots of common assumptions of ideas like 'national talent' and practical endeavours to identify and nurture this quality, this book explores the tension between the state's power to create identities that carry tremendous cultural capital and the citizen's appropriation of these possibilities to pursue their aspirations.

This is a tension that gained particular salience in the decade between 2010 and 2020, the period that this study took shape, first as a doctoral dissertation and then as a monograph. This period has witnessed an unprecedented and anguished questioning of the limits and relevance of the social formation of the 'nation'. Globally and nationally, this interrogation has been driven by a greater escalation and exposure of gender, race, and caste violence; the extremism and war triggering waves of refugee migration from various parts of the Arab world, Africa and Asia; the devastating human and social losses driven by increasing natural disasters and crises resulting from climate change, etc., all of which have been compounded by corruption, inefficiency, complicity, and brutality on the part of states across the globe. Simultaneously, in the Indian context, while the routine justification of state actions took the discursive form of 'nation-building', the form and responses to expressions of dissent, difference, anger, and fear have coalesced in a totalizing term – 'anti-national'. We have witnessed these tendencies being further consolidated during the overwhelming and paralysing distress unleashed in the wake of the eruption of the novel coronavirus pandemic across the globe in 2019 and 2020, with a resurgence of the term 'nation' as a unit of analysis and intervention (theories of the origin of the pandemic, control measures, restrictions on international travel, access to vaccines and medical aid, etc.).

The management of crises and disasters brings the state to the fore of one's attention, spotlighting successes and failures in a spectacular fashion. Yet there is a dichotomy to this perception of the state. On the one hand, the popular imagination is rife with the awareness of the corruption and indifference of central and state governments, which may be run by callous and inefficient political parties, with the support of a judiciary that does not always appear effective, an indifferent bureaucracy, and the police and military, whose violence can appear arbitrary. Yet, on the other hand, the same 'state' is perceived as different from these and continues to be 'a locus of desire' for citizens in its avatar as the provider of 'jobs, ration cards, educational places, security and cultural recognition'(Kaviraj, 2010, p. 30) The relationship between the opportunities for recognition and the pursuit of personal aspirations that is made possible by the modern state is consolidated in these quiet and mundane interfaces of state power and individual identities. Through exploring one such interface in the articulation of 'national talent', I invite you to reflect anew on what Sudipta Kaviraj calls 'the strange enchantment of the state'.

## Notes

1 Established in 1920, the CABE has functioned as the highest advisory body for the central and state governments in the field of education (except for the period from 1923 to 1935, when it was dissolved). Historically, it has been constituted by elected members from the Lok Sabha and the Rajya Sabha, and the representatives of the Government of India, State Governments and Union Territories and nominated members representing various interests like NGOs, academics, etc.

2 This book was an abridged version of an anthology produced by the Indian Academy of Sciences entitled *Lilavati's Daughters* (Godbole & Ramaswamy, 2008). The reference to Lilavati (the daughter of the twelfth century mathematician Bhaskara) reflects the content of the book, which contain essays by 100 women scientists of India.

## References

Adamsen, B. (2016). The Etymology of the Word "talent". In *Demystifying Talent Management* (pp. 77–87). Palgrave Macmillan UK.

Anderson, B. (1991). *Imagined Communities*. Verso.

Baral, B., & Das, J. (2004). Intelligence: What Is Indigenous to India and What Is Shared? In R. Sternberg (Ed.), *International Handbook of Intelligence* (pp. 270–301). Cambridge University Press.

Bhabha, H. (1994). DissemiNation: Time, Narrative and the Margins of the Modern Nation. In *The Location of Culture* (pp. 199–243). Routledge.

Bhattacharya, S. 2011. Molecular Biology. In Ramaswamy, Godbole and Dubey (eds). *The Girl's Guide to a Life in Science*. Young Zubaan.

Carson, J. (2007). *The Measure of Merit*. Princeton University Press.

Dalal, A. (2011). A Journey Back to the Roots: Psychology in India. In Cornelissen, M., Misra, G. & Varma, S. (Ed.), *Foundations of Indian Psychology: A Handbook*. Pearson.

Fairclough, N. (1989). Discourse, Common Sense and Ideology. In *Language and Power* (pp. 77–108). Longman.

Howe, M., Davidson, J., & Sloboda, J. (1998). Innate Talent: Reality or Myth. *Journal of Behavioural and Brain Sciences*, *21*(3), 399–419.

Jain, V. K. (2001). *An Appraisal of the NTSE-2001*. NCERT.

Kumar, K. (1986). Agricultural Modernization and Education. *Economic and Political Weekly*, *31*(35/37), 2371–2373.

Nambissan, G., & Batra, P. (1989). Equity and Excellence: Issues in Indian Education. *Social Scientist*, *17*(9/10), 56–73.

NCERT. (1964). *Annual Report of the NCERT 1963- 1964*. NCERT. (2006). *Annual Report of the NCERT 2005-2006*.

Oommen, T. K. (1989). India: "Brain Drain" or the Migration of Talent? *International Migration*, *27*(3), 411–425.

Pandey, S. S. (2004). *A study into the causes of poor participation and achievement of girl candidates of Uttar Pradesh at National Talent Search Examination*.

Philip, R. (2016). The Social Construction of Academic Ability: A Review of Literature. *Gyanodaya: Journal of Progressive Education*, *9*(2), 24–34.

Poulson, L. (1996). Accountability: A Key-Word in the Discourse of Educational Reform. *Journal of Education Policy*, *11*(5), 579–592.

Raina, M. K. (1984). *Background of the Talented*. Delhi. NCERT.

Raina, M. K. (1989). *An Intensive Study of the Accomplishments of National Science Talent Scholars*. NCERT.

Raina, M. K. (1991). *The Talented Scholars: Accomplishments and Worldview of the Talented*. *Delhi*: NCERT.

Ramaswamy, R., Godbole, R., and Dubey, M. eds., 2011. *The Girl's Guide to a Life in Science*. Young Zubaan.

Rudnikski, R. (2000). National/Provincial Gifted Education Policies: Present State, Future Possibilities. In Heller, K. A., Mönks, F. J., Subotnik, R., & Sternberg, R. J.

(Ed.), *International Handbook of Giftedness and Talent* (pp. 673–679). Elsevier Science Ltd.

Saxena, K. N. (1964). *A Report of the National Talent Search Scheme Examination. Delhi*: NCERT.

Saxena, K. N. (1967). *A Report of the Science Talent Search 1967.* Delhi: NCERT.

Schneider, A. L., & Ingram, H. (2008). Social Constructions in the Study of Public Policy. In *Handbook of Constructionist Research* (pp. 189–211). Guilford Press.

Thapar, R. (2009). The History Debate and School Textbooks in India: A Personal Memoir. *History Workshop Journal, 67*(1), 87–98.

Yagnamurthy, S. (2010). The Flight of Talent to Selective Academic Streams and Its Impact. In S. Kaur (Ed.), *Contemporary Issues in the Global Higher Education Marketplace: Prospects and Challenges* (pp. 73–86). Malaysia: National Higher Education Research Institute (IPPTN).

Zechariah, B. (2001). Uses of Scientific Argument: The Case of "Development" in India, c 1930-1950. *Economic and Political Weekly, 36*(39), 3689–3702.

# 2 'National Science Talent'
## Historicizing an Articulation

….. Three words are there: national, science, talent. I was national and I was talent. But science? If there were a simple national talent search, I would [be part of that]. Everything I did falls in that. I deserved that. Not a single rupee in that scholarship went waste…. I was an outstanding national scholar and I still am today. But I am a journalist.

This was how James, a scholar selected through the National Science Talent Search (NSTS) in 1968, reminisced about the impact of the programme on his life. To contextualize and interpret complex representations of talent such as this – where one's ability and its recognition and the assessment of one's personal and professional trajectory are layered by emotions, expectations, interrogation, and reflection – one must seek the points of intersection between biography and history.

This chapter hopes to set in place some markers that allow us to make sense of the discourses that were articulated together in the inception of the National Science Talent Search Examination. To do so, I review literature that allows us to trace two discursive movements across the history of ideas. The first is the transformation of the meanings of the English word 'talent' and its articulation with the discourses of the individual and the nation during the seminal social transformations witnessed between the seventeenth and nineteenth centuries in the Euro-American context. This entails tracing some internal developments of the word 'talent' in the English language and how these allow it to be embedded within a certain relationship between the individual, state, and the nation, which developed from the eighteenth century onwards in Europe and North America. The power of this discursive formation lay in its demand that forms of social organization must consider the ideal conditions for human flourishing.

The subsequent section explores how the conceptualizations of individual ability such as 'talent' and 'intelligence' were shaped by ideas of nationalism and colonialism from the eighteenth century. This then provided a direction for the emerging social and behavioural sciences in the nineteenth century. In the first half of the twentieth century, the demands of the World

DOI: 10.4324/9781003344902-2

War and the race for technological superiority set in place the context for the model of the state-sponsored science talent search in the project of nation-building.

When these discourses filtered into the Indian context in the nineteenth century, traditional conceptions of ability had to contend with new conceptions of the individual that emerged during the colonial period. In the decades before 1947, the modern promise of 'individual development' grated against the colonial identity of being a 'subject' and played an important role in shaping anti-imperialist sentiment, and, subsequently, various incipient 'nationalisms' as a response to it. In this context, Jawaharlal Nehru was instrumental in promoting a discourse of 'nationalism', which was based on 'scientific development' in his writings as well as his administrative and policy decisions as the independent nation's first Prime Minister. The burden of achieving the promise of science for providing solutions for the country's socio-economic problems was laid on the back of 'experts', who were meant to provide leadership in this direction for the new nation. The final part of the chapter moves on to examine the impact of this direction taken during the first two decades after Independence in the way that the Indian state has been conceptualized, as well as its role in nation-building. The chapter concludes by placing the relationship between the discourse of 'talent' and the 'nation' as embodied in an examination like the National Talent Search Examination (NTSE) against the background of the history of ideas, which have been traced here.

## The Discourse of 'Talent' and the 'Individual' in the Euro-American Context

The English word 'talent' traces its root to the Greek word 'talenton'. This was a unit of weight that was used by Assyrians, Babylonians, Greeks, Romans, and other ancient people groups, thereby denoting something concrete and measurable. It was then transformed into a monetary unit when a value was attributed to one talent of silver (Tansley, 2011). The word 'talent' enters the English language through translations of the Bible, where the word was used to denote certain quantities of money. In the Middle Ages, the term began to undergo a series of transformations related to the interpretations and translations of the Parable of Talents in the Gospel of Matthew (chapter 25, verses 14 to 30) (Tansley, 2011).

In the parable, 'talent' is used as a noun with a denotative meaning of a precisely defined denomination of silver (Adamsen, 2014, p. 78). A man distributes to three of his servants 'to every man according to his several ability' five talents, two talents, and one talent, respectively (King James Version, Matthew 25.15) before going on a journey. When the man returns and settles accounts, the servant who increased his five talents to ten and the servant who multiplied two talents to four are rewarded. The servant who received only one talent and who 'hid [his] in the earth' is rebuked by the master who says '*Thou oughtest therefore to have put my money to the exchangers, and then at my coming, I should have received mine own with usury'* (Matthew 25. 27). As the servant's punishment, his talent is given to the one who had accumulated ten talents.

> For unto every one that hath shall be given, and he shall have abundance; but from him that hath not shall be taken away even that which he hath. And cast ye the unprofitable servant into outer darkness: there shall be weeping and gnashing of teeth
>
> (Matthew 25. 29–30)

In the parable, talent as a definite economic value is enmeshed in a moral economy between the master and servant. The themes of endowment, responsibility, accountability, profitability, merit, and waste are also intended to provoke rumination of the relationship of the individual and God.

By the fourteenth century, we see the emergence of a new adjectival function of the word 'talent' in English. Its single denotative meaning was replaced by 'more abstract, indefinite and indefinable meanings such as 'inclination', 'disposition', 'mental endowment' and 'natural ability' (Adamsen, 2016, p. 78). These abstractions allowed the word 'talent' to also describe the state of being talented, while simultaneously referencing that this state is 'something given' by God or nature. In the process, it lost its relation to concrete objects and began to be embodied within the individual (Tansley, 2011). By the seventeenth and eighteenth centuries, 'talent' as both noun and adjective began to point to 'aptitudes and faculties of various kinds (mental orders of a superior order, mental power or abilities)' (Tansley, 2011, p. 267). That this transformation influenced the translation of 'talent' into other European languages (German, Russian, French, Danish, and Polish) is evident from its representation in their dictionaries, which take the position that talent is the manifestation of innate giftedness in a field of endeavour, and it is linked to outstanding performance in some way. This kind of Anglo-European construction of talent is different from non-European usages (Holden & Tansley, 2007).

Adamsen (2014) argues that its transformation of the meaning of talent from a monetary unit to more abstract ones such as 'inclination', 'disposition', 'mental endowment', 'natural ability', etc. is a process by which it became an *'accidental designator'* (i.e., a word is independent of the object it refers to in the real world) (Kripke, 1980) and a *'floating signifier'* (a term that can point to a multiplicity of meanings or an indefinable meaning) (Levi-Strauss, 1968; Mauss, 1968). This ambiguity is a legacy that the word 'talent' continues to carry today, and for certain fields, such as 'the science of talent management', this is a problem (Adamsen, 2014). But the development of such an internal ambiguity in the word is also what allows it to become articulated with other discourses, especially that of individualism and nationalism, in interesting ways.

The internal transformation of the word 'talent' to include an adjectival sense primed it to be embedded within the discourse of individualism that emerged in the eighteenth century. The heart of the modern liberal perspective on the individual is shaped during this period, and it was a relatively new idea that every individual has a particular way of being true to one's self, which has to be discovered by him or her within themselves. The work of philosophers like Jean-Jacques Rousseau was central to establishing this inward turn of modern philosophical thought by imagining modes of social organization which would create conditions for the flourishing of the potential of every individual (Taylor, 2009). The power of this discourse of individualism is evident in the fact that it formed the basis of the revolutionary challenge to the primacy of a social order based on birth, privilege, religious affiliation, and communal identities in eighteenth-century Europe (France to be specific) and in colonial North America. The redefinition of the basis of individual identity was also central to the emergence of a new type of group consciousness during this period, which was very different from traditional affiliations. This perspective is perhaps most famously articulated by Benedict Anderson in *Imagined Communities* (1991). In the case of Europe, Anderson argues that the emergence of print technology and cheap mass-produced 'print commodities' vastly expanded the number of people who believed that they shared the same common stories. The experience of reading newspapers, novels, etc. created a sense of sharing simultaneity and contemporaneity ('a homogenous empty time') with a vast number of unknown individuals. This simultaneity and contemporaneity evoked a shared cultural history. This group consciousness was the beginning of an 'incipient nationalism' that would later transform the political order in Europe and North America in the eighteenth and nineteenth centuries (Anderson, 1991). Once established, this connection spilled over to the twentieth century and found a place in the competing projects of nation-building that gave shape to seminal wars and new distributions of global political and economic power.

Carson (2007) argues that Enlightenment authors of the eighteenth century used the word very carefully and always in the plural ('talents'), connecting it to other terms like 'abilities', 'capacities', and 'faculties'. Taken together, the emphasis in this vocabulary was that individuals possessed these attributes in different degrees. Common to our engagements with these terms even today, there was considerable disagreement as to the nature of these differences, their distribution among the populace, and their origins (education or 'nature', the latter being further segmented into heredity or chance). The most important significance of this terminology, especially that of 'talents', was that it provided a 'natural' criterion for delineating and discussing human differences. 'Talents' was one way of speaking like a democrat and yet still being able to justify inequalities' (Carson, 2007, p. 16). For example, American republican leaders like John Adams and Thomas Jefferson, despite their differing opinions on several issues, held that positions of influence in government and society should be or would inevitably be filled by those with superior abilities. The eighteenth-century speculations about human nature had a tremendous impact upon the republican project because it attempted to shift the eligibility for leadership away from the vagaries of birth to cultural, scientific, and economic accomplishments. In doing so, a direction was provided for the nascent human sciences in envisioning 'social formations consonant with the developing notions of the republican citizen, the enlightened society and the self-interested economic actor' (Carson, 2007, p.16).

The articulation of talent with other discourses of ability occurs over the nineteenth century through the work performed by the nascent social and behavioural sciences, which contended that individual differences in cognitive potential could be 'scientifically' quantified. The construction of 'intelligence' and its connection to colonial conceptions of race has already been extensively researched. For example, Stephen Gould's *The Mismeasure of Man* (1981) draws our attention to the historical processes and consequent implications of the abstraction of just one aspect of the human brain, i.e. cognition, into the idea of 'intelligence'. The discourse of intelligence and its technologies of measurement was used to produce data that ratified the categorization of human beings along racial lines. His extensive reanalysis of the data (such as skull measurements) by certain craniometrists and polygenists of the seventeenth and eighteenth centuries reveals how the belief in the racial superiority of white Europeans led to (intentional) errors in the measurement of data at times. These ideas shaped perspectives of how intelligence was inherited and distributed in predetermined ways among racial populations. They were also used by colonial administrators in the statecraft involved in the governance of their territories (Gould, 1981).

The lesson that we can draw for our exploration of the relationship between talent and nation-building from the work done on 'intelligence' is to be aware of how the basis of a country's nationalist self-conception shapes other discourses appropriated by it. Carson (2007) reminds us that the reception of the burgeoning racial sciences of the late nineteenth century and the application of the scientism of psychology in various domains in the twentieth century were not the same across countries. He makes this argument in the cases of the United States of America and France, both of which (as discussed before) was founded on the ideal of the democratic republic. American nationalism is crucially tied to the principle of individualism emanating from the discourse of 'natural rights'. In contrast, French nationalism invokes the nation as the embodiment of the general will of the people and therefore, the safeguard of the larger social good. These differences in the American and French notions of meritocracy both constructed fundamentally different educational systems and shaped their reception of the behavioural sciences. For example, the technology of intelligence testing was adapted to different ends in both countries. Since members of privileged socio-economic groups generally scored well on intelligence tests, the concept of 'intelligence' contributed to the preservation of the American social hierarchy while allowing room for exceptional members of historically marginalized groups. In France, by contrast, the educational system was the primary gatekeeper for entrance into the tech-nocratic elite. Intelligence tests and the institutional and cultural roles of such identification retained Alfred Binet and Theodore Simon's original intent, as being associated with the identification of the mentally deficient rather than the skilled well into the 1930s. Intelligence was considered as an aid to individual self-understanding and as a tool for teachers to under-stand their students better (Philip, 2016), rather than a unitary category into which mental differences between people could be collapsed as in the American case.

A more specific exploration of the relationship between 'talent' and 'nationalism' in the twentieth century is found in Sevan Terzian's *Science Education and Citizenship: Fairs, Clubs and Talent Searches for American Youth, 1918–1958* (2013). The history of the Westinghouse Talent Search that Terzian sketches is particularly of interest to us because of how India's science talent search was inspired by it. He explores how the World Wars and the interregnum significantly shaped the United States' approach to science education, including the idea that talented students in science must be located through a talent search and nurtured to promote national secu-rity. A country's superiority in war was understood to be directly propor-tional to its ability to produce good scientists. This is contextualized through his analysis of the institution and significance of the Westinghouse (now Intel) Talent Search Examination from its inception during the Second World War. In the 1920s and 1930s, the progressivist ideas of John Dewey spurred American science educators (like Morris Meister) to create plat-forms like science fairs and clubs to prompt children to investigate their natural and social environment. By the late 1930s, the onset of the World War saw a political and economic consensus that created an unprecedented emphasis on grooming high-achieving youth with intellectual capital' for national economic and military benefits. This perspective rivalled long-standing progressive efforts to promote widespread scientific literacy. Terzian pinpoints the sponsorship of New York's World Fair of 1939 by the Westinghouse Corporation and the disagreements between various interest groups in how the research of the students was to be selected and presented as a pivotal moment of transition. The nationwide Westinghouse Science Talent Search was instituted in 1942 against the backdrop of the mobiliza-tion of schools and science clubs for national defence. The predominant rationale for this was that the scientific acumen of the intellectually gifted students would fortify the US militarily and materially in the wartime. In the subsequent atomic age and Cold War, science fairs, clubs, and talent searches maintained their meritocratic raison-d'etre as serving national defence and domestic prosperity. Terzian extends his analysis to the pres-ent-day context of the National Academies of Science and the National Science Board of the United States, which exemplify the continuing ten-dency to enlist science education to bolster the country's economy and security, rather than emphasizing its possibilities for global participatory democracy (Terzian, 2013).

## 'Talent' and the 'Individual' in India: New Connections Bridging Tradition and Modernity

The relationship between nationalism and the discourse of talent has taken a very different trajectory in the Indian context. India has had a long cultural tradition that has grappled with the idea of how intellectual and creative powers manifest beyond the range of the 'typical' or the 'normal'. These predate the modern Indian nation and even the modern ideas of 'individualism'. In the Indian context, it must be remembered that the individual was not the 'basic social unit' of the traditional social order before the colonial period nor was individual atomism considered desirable. In *The Inner World* (1981), Kakar argues that the traditional worldview of Hinduism was explicitly oriented towards the fusion of the self with others/The Other. The fundamental ideal of *mokṣa*, conceptualized as the aim of one's existence, reflected the cultural-psychological predominance of 'primary processes of thinking, communal relationships and otherworldly realities' (Kakar, 1981, pp. 104–105). Consequently, there was a strong tendency to resist the values of individualism. Ideas of personal flourishing, the worthwhile life, notions of success, social contribution, and achievement (which were rooted in a predominantly upper-caste Hindu world image and which focused on how the goal of enlightenment and release from rebirth may be attained) were significantly different from those underlying words like 'talent' and 'intelligence' in the Euro-American context.

Let us consider the word 'pratibhā', which is used as a translation for the word 'talent' in several Indian languages, including Hindi. (It is also used in the translation of the NTSE as the Rashtriya Pratibhā Khoj Pareeksha). In classical Sanskrit, *pratibhā* means 'a flash of light or revelation' (Kaviraj, 1968). It refers to a supersensuous apperception that is distinct from intellect, and such knowledge is characterized by tremendous clarity and immediacy (Kaviraj, 1968, p. 2) and is deemed to be the source of the excellence and creativity displayed in outstanding poetic and artistic creation (Shulman, 2008). The possession of 'pratibhā' is represented as transcending even the contributions that an individual makes to society, which is often how the modern achieving or successful individual is defined. In classical Hindu and Buddhist literature, pratibhā is the highest state of mystic consciousness before kaivalya or mokṣa, which is the liberation from the cycle of birth and death (Kaviraj, 1968). Within the Hindu world image, where the attainment of 'mokṣa' is the ultimate aim of human existence (puruṣārtha), it is the liberated individual who is the cultural ideal of this worldview, i.e. the one who has overcome ignorance or false consciousness (avidyā) (Kakar, 1981). Therefore, the term 'pratibhā' points to an immediately recognizable excellence and virtue, the promotion of which was considered among the highest aims of human existence in this worldview.

In common with other Asian traditions of conceptualizing the possession of exceptional knowledge, creativity, and intellectual prowess (Holden and Tansley, 2007), pratibhā is traced to an individual's discipline and tutelage, rather than genetic endowment. While the earliest Vedic literature does not throw direct light on the nature of learning processes, by the late Vedic period, it is established that 'the correct cadence, the correct intonation, the correct accent, these can only be learnt direct from a teacher, it cannot be accomplished from books' (Shrimali, 2011, p. 4). Therefore, it is not surprising that the attainment of pratibhā is tied to intense discipline and effort under the tutelage of a guru. In his essay, 'The Doctrine of Pratibhā in Indian Philosophy', Sanskrit philosopher Kaviraj describes the stimulating role attributed to the influx of the teacher's spiritual energy in awakening the student's latent power or inner potential. It is held that this 'infusion of energy' purifies the soul of the disciple and creates a state of concentrated mind (cittaśuddhi). The completion of this process is associated with a physical manifestation of pratibhā as a luminous eye in the middle of the forehead, i.e. the Divyacaksuh/the third Eye of Śiva/the Eye of Wisdom/the Eye of the ṛṣi. Since this eye is opened by the grace of the guru, the latter is usually called the 'giver of the eye' (Kaviraj, 1968).

Simultaneously, it is important not to lose sight of the fact that the cultural ideal of 'talent' was theoretically conceptualized in a manner that tended to preserve patriarchal and Brahminical hegemony over the social order. Despite the many illustrious names (Apālā, Ghoṣā, Lopāmudrā, Sikatā, Viśvavarā, Gārgī, Maitreyī, etc.) that are highlighted to eulogize the state of women's education in the times when the Vedic texts were composed, the historian Shrimali advises caution in extrapolating from these cases. These women largely belonged to the upper classes and were also exceptions even in their times. Additionally, they could not go beyond a point in acquiring knowledge (Shrimali, 2011, p.7). The structure of disseminating knowledge seems to have become further closed, especially from the late Vedic period onwards, when women and those who were not of the twice-born castes were not instructed in Sanskrit. Additionally, several classical canonical texts contain injunctions and penalties against them reading these (Thorat & Kumar, 2008). The words and sentences of the Sanskrit language were given mystical importance, in that they were assumed to be capable of illuminating not just the phenomenal world but also the reality behind it. The perfect knowledge of Sanskrit was elevated as being essential to obtaining merit (dharma) and attaining spiritual elevation (abhyudaya) (Scharfe, 2002). In such a context, the ideal of pratibhā would be restricted very narrowly as applicable to upper-caste males.

A didactic example of the restriction of the access to education and training given to talented individuals outside 'the twice-born' castes is found in the Mahābhārata. The young Ekalavya, who was a member of the forest tribe of the Bhils, greatly desired to master the skill of archery. Knowing that his birth precluded him from being a student of the famous teacher Droṇa, the boy made his statue instead and followed a program of disciplined self-study before it, adopting him as a guru. On witnessing Ekalavya's potential of surpassing the skills of his royal students (including his favourite, Arjuna) and learning of the boy's symbolic adoption of himself as his teacher, Droṇa demands the boy's right thumb as 'guru dakṣiṇā (payment on completion of a course of study). In this story, the teacher is motivated by the desire to preserve the status quo and to eliminate any challenge to this order (Shrimali, 2011, p. 6). The traditional narrative adds the dutiful acceptance of this lot by Ekalavya, who is presented as having internalized the brutal reality of caste-based access to knowledge.

The colonial encounter created a paradigm shift in terms of how the individual and the development of his or her potential were conceptualized. The institutions of the state created by the British created space for a different discourse of individualism. New English institutions like the courts, the civil service, the university, schools, etc., along with new conceptions of property (like land rights) and profession, created possibilities for self-development, distinction, and mobility based on the idea of individual merit (Beteille, 1983). These changes were not sudden because the British did not conquer India in one fell swoop. In contrast, many of them were responses to the demands of administration and were tied to issues such as 'mapping territory and enumerative practices for the levy of taxation, the creation of an army for territorial control, the gradual exposure of a limited native segment to modern European education, etc' (Kumar, 2005). The transformations brought by the colonial state were embraced, not just by the upper castes but also by an elite among the lower castes. They gained concrete benefits from this opening up of education and employment (Omvedt, 2004).

As we discussed in the first chapter, it was the influence of the reforms undertaken by the British in education from 1854 onwards that began to change native conceptualizations of pupil ability and worth. The institution of a new mechanism of examinations linked education with the competition for the achievement of status and power, such as that embodied by the civil service (Kumar, 1988). Rather than subject competence, these exams tested individuals for their mastery over prescribed textbook content. However, examination success increasingly began to function as a proxy for academic 'talent' and 'merit'. The discourse of psychometry was marginal to educational processes in nineteenth-century India. However, conceptions like 'intelligence' found purchase in urban British medical practice in the nineteenth and twentieth centuries. The discourse of 'intelligence' provided a vocabulary that was used to meet the various administrative demands of colonialism, such as settling property and succession issues, categorizing inmates at lunatic asylums, and differentiating between forms of disability (Miles, 1998). With the commencement of teaching psychology as a discipline at the University of Calcutta in 1916, mental testing with a view to understanding the developmental trajectory of children took hold in the Indian context in the 1920s and the 1930s (Dalal, 2011) (Thorat & Kumar, 2008). As in the case of France, the discourse of intelligence tended to be applied more for marginal attempts to address the needs of the exceptions on either side of the normal distribution curve, while the institutional educational system was treated as a valid means of attesting to pupil ability, a certain 'schoolhouse giftedness' at least (Miles, 1998).

The link between individualism and nationalism was also forged under the colonial experience. The promise of 'individual development' offered by 'modernity' could not be reconciled with the colonial identity of being a 'subject'. This tension played a crucial role in shaping anti-imperialist sentiment and various incipient 'nationalisms' as a response to it. The opportunity which the anticolonial struggle offered with respect to imagining the identity of the nation was taken up by a diverse range of stakeholders. Prescient national leaders, be it Gandhi, Ambedkar, or Nehru, had realized that the colonial transformation of the country created space for a language of 'nationalism', which could treat India not as a primordial reality but as a project which had to be developed. This was an opportunity to imagine the kind of social organization which could ameliorate inequalities and unleash the potential of the people who had been suppressed both by culture and colonization. Additionally, it was necessary to craft a conscious response to various competing cultural-revivalist nationalisms, which were based on a perceived threat to religious identity and the possible erosion of the privileges of caste hierarchy and patriarchy. Such projects also channelled the logic of religious majorities and minorities as well as the arithmetic of perceived scores to be settled after centuries of Muslim dominance. For example, the rejection of a hierarchical coexistence of the secular and the religious in the imagination of the 'nation' was captured in the respective demands for 'Akhand Hindustan' and 'Pakistan' (Kumar, 2007b).

## Articulating 'Science' and the 'Nation'

It is within such a context that Jawaharlal Nehru played a crucial role in consolidating a narrative for the Indian nation based on the liberal ideology of modernization and development, both in his writings and in his administrative and policy decisions as India's first Prime Minister. Within this narrative, space was carved out for an important discourse of scientific expertise and the development of the capacity of select individuals to provide the necessary leadership in this direction for the new nation. For example, in *The Discovery of India*, which was first published in 1946, Nehru attempts to demonstrate that what unifies people of diverse backgrounds, regions, and identities into members of the entity 'India' is a shared historical trajectory. It connects the past ('*a cultural tradition through five thousand years of history, of invasion and upheaval, a tradition which was widespread among the masses and powerfully influenced them*'), the present reality of colonial subjection and the future ('*one of intimate co-operation, politically, economically and culturally between India and the other countries of the world*') (Nehru, 1989, pp. 50, 52). He compares this trajectory of India's with that of Euro-American societies. The lesson that is drawn is that periods of socio-economic stagnation can be overcome, and vitality can be regained through a conscious reorganization of their economy, polity, and society in accordance with the demands of modern science.

In other words, it was science that catalysed all the other necessary developments for a successful nation-state.

Nehru's understanding of the 'scientific approach' was a nuanced one. He was aware that the 'scientific method of objective inquiry' had much beyond its scope, including the provenance of art, poetry, aesthetic appreciation, or even moral categories such as goodness. He also did not hold that the full potential of 'the real temper of science' had been reached in the Western world despite its use and celebration of the material and social contributions of science (Parekh, 1991). However, in common with the leaders of the post-Revolution French state, Nehru saw in the scientific approach the means through which citizens would learn to be sovereign and to be associated with the state apparatus in a free India. Like the French leaders of the First Republic, he believed that scientific institutions would be instrumental in creating a scientific public, which in turn, would create 'necessary future transformations such as the full realization of human rational potential' (Harrison & Johnson, 1994). In the *Discovery of India*, he described the disposition of such a citizenry as oriented by

> the adventurous and yet critical temper of science, the search for truth and new knowledge, the refusal to accept anything without testing and trial, the capacity to change previous conclusions in the face of new evidence, the reliance on observed fact and not on pre-conceived theory, the hard discipline of the mind- All this is necessary, not merely for the application of science but for life itself and the solution of its many problems.
>
> (Nehru, 1989, p. 512)

This idea of Nehru, which links science and the solution of social problems, was part of a discourse of 'development' that had evolved in nineteenth-century Europe. There were several theoretical as well as practical attempts to explore how social change could be produced or directed at will through the use of scientific knowledge. The emergence of a discourse of 'planning' was central to this shift. In the nineteenth century, urban planning came into its own in Europe. Its emergence was tied to the demands of meeting the unprecedented challenges posed by the Industrial Revolution concerning poverty, unemployment, health, hygiene, education, and so on (Escobar, 2010). The solution to these challenges was sought through constructing them as 'social problems', which required detailed scientific knowledge about the populace. This knowledge was then to be used as the basis for suitable interventions. The aim of governance became the 'efficient management and disciplining of the population to ensure its welfare and good order' (Escobar, 2010, p.146). This idea of 'scientific development' of a society was also linked to a parallel discourse which emerged contemporaneously and which 'invented' the 'economy' as 'an independent domain apparently separated from morality, politics and culture' (Escobar, 2010, p.147). This discourse was shaped by the convergence of the insights of classical political economy and the philosophies of utilitarianism and individualism in the nineteenth century. This was coupled with the spread and institutionalization of the market, especially because of Europe's colonial ambitions. This 'invention' of the economy was central to the way that the State was imagined. Its role in the management of resources through 'planning' added the connotation of 'welfare' to the discourse of 'development' (Escobar, 2010). Additionally, the discourse of 'scientific planning' had gained greater visibility as a principle for social action in the 1920s and 1930s through its use and advocacy in diverse contexts such as 'the mobilization of national production during World War I, Soviet planning, the scientific management movement in the US and Keynesian economic policy' (Escobar, 2010).

This 'scientific' and 'economic' discourse of 'development', which left such a deep imprint in Nehru's writings as well as his political and administrative decisions, had begun to play a greater role in articulations of Indian nationalism from the late nineteenth century onwards (Deshpande, 2003). Over the first decades of the twentieth century, the economic idiom of analysing the 'nation', its history, and its future provided a new vocabulary for imagining the national unity of India. The economic discourse of 'nationalism' enmeshed together several ideas to form a relatively coherent narrative over the course of the Independence movement. This included the imagined 'fabulous wealth of India' prior to the British rule; the colonial exploitation, which drained these riches; the language of indigenous production and consumption in the 'swadeshi movement'; and finally the technical expertise of socialist planning and development economics, which could anticipate the nation's progressive fulfilment of its 'potential' (Deshpande, 2003). As we briefly touched upon in the first chapter, in the 1930s, a confluence of factors like the cumulative experience of the economic Depression, the introduction of provincial autonomy in 1935, and the growing anticipation of Indian independence demanded a re-examination of how the economy and polity should be organized to usher national prosperity. Its catchphrase 'national reconstruction' was an umbrella which brought together visions of change and growth in various sectors of the economy under the leadership of technically qualified experts (Zachariah, 2001). Deepak Kumar (2007a) notes that the National Planning Committee, which was constituted by the Indian National Congress in 1938, reflected this impulse. Under the leadership of Jawaharlal Nehru, 29 'expert' subcommittees were formed to address different aspects of 'national reconstruction', including agriculture, industries, population, labour, education, health, energy, irrigation, afforestation, communication and housing (Kumar, 2007a, p. 112). This ascendance of 'expertise', and in particular 'scientific expertise', would continue into the first two decades after independence, where such individuals played key parts in the formulation of public policy, supported by Nehru's endorsement of and commitment to science-led developmentalism.

The prominence given to expert personnel in the discourse of planning cast scientists as a highly influential group in a new India. Before

Independence, India had several scientists who had achieved an international reputation for excellence through the prizes they won (including the Nobel Prize) and the elite fellowships they held such as that of the British Royal Society. The recognition on a world stage achieved by Indian scientists such as Prafulla Chandra Ray, Jagdis Chandra Bose, Satyendranath Bose (though it was later), Meghnad Saha, or C. V. Raman was the source of much national pride and inspiration. Their individual access to powerful international networks was also an asset for a nascent nation, in the inevitable demands made on India regarding international negotiations and transactions after Independence (Anderson, 2010). Even before 1947, such scientists were aware of several strategic developments of their times and their implications for India, such as the important discoveries about nuclear fission in 1939 and the potential of atomic energy as a source for an energy-deficient nation (Anderson, 2010). A heightened public interest in scientific practice (which was also expressed in the mushrooming of scientific journals and greater space given in periodicals and magazines for scientific inventions and discoveries) also encouraged several scientists to take their 'expert' role seriously and to advocate the importance of science to a wide and constantly increasing readership among the middle class (Zachariah, 2001, p. 3690). In many ways, such interchanges contributed to a particular mythicized image of the scientist.

Both Zachariah (2001) and Anderson (2010) highlight how this cultural prominence of the scientist was a relatively new phenomenon, because technology and technical education leading to practical applications of engineering tended to receive far more support in the colonial context than fundamental research in the basic sciences. Nair et al (1997) compare such representations with that of the engineer to argue that 'unlike the pure scientist, myths of troublesome and heroic discoveries seldom attach to the engineer. How many technologists, whatever their contribution, have the towering intellectual reputation that attaches to an Einstein, a Raman or a Hawking?' (Nair et al., 1997, p. 8). By the mid-twentieth century, there was a growing prestige around 'science' as a professional career path, where one's abilities would be used in the service of the nation.

## 'National' 'Science' 'Talent'

The aim of the genealogy of ideas, which has been presented in this chapter, has been to help us understand the discourse represented by 'national talent' – a category specifically created through the policies and practices of the state to identify individuals who are tasked with aim of building the 'nation' through professional contributions. We come back to the theme with which this chapter commenced, i.e. the importance of language in social practice and history. To consolidate some themes in the historical development of the relationship between talent and nation-building as well as to set the tone for the chapters that are to follow, I would like to draw attention to a concept developed by the political theorist Murray Edelman

(1964, 1977), which has been fruitfully used by researchers working in the area of education politics ((Troyna, 1994) (Poulson, 1996)).

Edelman identifies the importance of 'the condensation symbol' in symbolic political language. Such a symbol is able to 'condense into one symbolic event, sign or act; patriotic pride, anxieties, remembrances of past glories or humiliation, promises of future greatness; one of these or all of them' and thus evoke the emotions associated with particular situations ((Edelman, 1964, p. 6) cited in Poulson, 1996, p. 581). At one level, the condensation symbol works very well because the elements within it have no clear referential relationship and therefore, it may represent several states or emotions (Poulson, 1996). The idea of 'national science talent' is also a powerful condensation symbol, drawing into its ambit a certain moral power and promise that have accreted to it from various historical contexts and social traditions.

The discourse of the 'nation' adds considerable gravitas to conceptions of 'individual talent', when they are coupled. This is because not only is the nation comprised of individual citizens, but the nation can also be represented to external and internal audiences by individuals in such a way that the nation and the individual are one (Verderey, 1993). This association between the nation and the individual is often capitalized in politics (e.g. the slogan of yore 'Indira is India and India is Indira'), in international sporting events, beauty pageants (Miss India, etc.), the achievements of scientists (who may just be of Indian origin and not even necessarily holders of Indian citizenship), and so on in a variety of fields. At this juncture, it is important to note that individuals do not automatically think or act 'nationally', i.e., as a 'member of the nation'. Rather they are socialized into such thought and behaviour patterns through the continuous work performed by the institutions of the modern state in creating and sustaining a national consciousness among the members of its body politic. In this context, we must remember that a critical aspect of this work of the state is its temporal dimension. The official narrative of the 'nation' upheld by the state tends to be split into what Bhabha (1994) calls a 'double time'. On the one hand, there is a sense of the 'nation' being in the making, progressing towards some final destiny. This requires that 'the people' be continuously treated as the 'object of a national pedagogy' so that they can play their part in fulfilling it. On the other hand, the unity of the people and their permanent identification with the nation has to be continuously affirmed as something that has already been accomplished. This affirmation must take on several symbolic forms of signification, repetition, and performance (Bhabha, 1994).

The significance of choosing the NTSE for a closer study in this research, rather than any other competitive exam, lies in the fact that it is the only national-level test conducted for school students under the aegis of the Ministry of Education and the NCERT. Therefore, the discourse of talent within this examination is closely connected to the larger policy perspective of the country on education. This is important because it is at the level of policy that the 'talented student' is first constructed as a special category among students who require special support from the state (Schneider & Ingram, 2008). Examining this construction in policy allows us to explore the state's ability to create new identities which possess tremendous symbolic capital. This capital is built on the perceived oneness between the state and the nation. To be recognized by the state is to be recognized 'nationally', and this is a source of great prestige.

The discourse of 'national talent' generated and sustained by the NTSE is embedded in a complex matrix of *institutions* (the Ministry of Education; the erstwhile Planning Commission; the National and State Councils for Education, Research and Training (i.e. NCERT and SCERTs); Central Advisory Board of Education (CABE); various state examination bodies, universities, schools, coaching centres, and media such as television, newspapers, and internet), *individuals* (students, bureaucrats, administrators, politicians, psychologists, educators, coaches/tutors) and *material infrastructure and objects* (buildings, offices, examination centres, computers, software, documents, etc.). While this book will not touch upon all of these aspects, in the subsequent chapters, I explore some dimensions of this through the discourse of 'talent' in education policy, the institutional shape that the talent search took, and the experience of this identity by scholars selected through the talent search.

## References

Adamsen, B. (2014). Do We Really Know What the Term "talent" in Talent Management Means? And What Could be the Consequences of Not Knowing? *Philosophy of Management*, *13*, 3–20.

Adamsen, B. (2016). The Etymology of the word "talent". In *Demystifying Talent Management* (pp. 77–87). Palgrave Macmillan UK.

Anderson, B. (1991). *Imagined Communities*. Verso.

Anderson, R. (2010). *Nucleus and Nation*. Chicago University Press.

Baral, B., & Das, J. (2004). Intelligence: What is Indigenous to India and What Is Shared? In R. Sternberg (Ed.), *International Handbook of Intelligence* (pp. 270–301). Cambridge University Press.

Beteille, A. (1983). Equal Distribution of Benefits. In *Idea of Natural Inequality and other Essays* (pp. 168–197). Oxford University Press.

Bhabha, H. (1994). DissemiNation: Time, Narrative and the Margins of the modern nation. In *The Location of Culture* (pp. 199–243). Routledge.

Carson, J. (2007). *The Measure of Merit*. Princeton University Press.

Dalal, A. (2011). A Journey Back to the Roots: Psychology in India. In Cornelissen, M., Misra, G., & Varma, S (Ed.), *Foundations of Indian Psychology: A Handbook*. Pearson.

Deshpande, S. (2003). *Contemporary India*. Viking.

Edelman, M. (1964). *The Symbolic Uses of Politics*. University of Illinois Press.

Edelman, M. (1977). *Political Language: Words that Succeed and Policies that Fail*. Academic Press.

Escobar, A. (2010). Planning. In W. Sachs (Ed.), *The Development Dictionary: A Guide to Knowledge as Power* (2nd ed., pp. 145–159). Zed Books.

Gould, S. (1981). *The Mismeasure of Man*. Norton.

Harrison, C., & Johnson, A. (1994). Introduction: Science and National Identity. *Osiris*, 24(1), 1–14.

Holden, N., & Tansley, C. (2007). Talent in European languages: Philological analysis reveals semantic confusions in management discourse. *Critical Management Studies Conference, Manchester Business School*.

Kakar, S. (1981). *The Inner World*. Oxford University Press.

Kaviraj, G. (1968). The Doctrine of Pratibha. In *Aspects of Indian Thought* (pp. 1–44). The University of Burdwan. https://archive.org/details/Aspects.of.Indian.Thought. by.Gopinath.Kaviraj/page/n1/mode/2up

Kaviraj, S. (2010). On the Enchantment of the State: Indian Thought on the Role of the State in the Narrative of Modernity. In *Trajectories of the Indian State*. Permanent Black.

King James Bible. (n.d.). King James Bible Online. https://www.kingjamesbibleonline. org/ (Original work published 1769)

Kripke, S. A. (1980). *Naming and Necessity*. Basil Blackwell – Oxford.

Kumar, K. (1986). Agricultural Modernization and Education. *Economic and Political Weekly*, 31(35/37), 2371–2373.

Kumar, K. (1988). Origins of India's Textbook Culture. *Comparative Education Review*, 32(4), 452–464.

Kumar, K. (2005). *The Political Agenda of Education*. Sage.

Kumar, D. (2007a). Colony and Science: A Study of British India. In *Science, Technology, Imperialism, and War* (pp. 89–120). Pearson Longman.

Kumar, K. (2007b). *Battle for Peace*. Penguin.

Kumar, P. S., & Chatterji, D. (2009). A Decade of KVPY- A Challenging Experience. *Current Science*, 97(10), 1282–1286.

Levi-Strauss, C. (1968). Introduction a l'ceuvre de Marcel Mauss. In *Sociologie et anthroplogie* (Quatrième, pp. 41–43). Les Presses universitaires de France.

Mauss, M. (1968). *Sociologie et anthroplogie*. Les Presses universitaires de France.

Miles, M. (1998). Professional and Family Responses to Mental Retardation in East Bengal and Bangladesh, 1770s-1990s. *International Journal of Educational Development*, 18(6), 487–499.

Nair, R., Bajaj, R. & Meattle, A. (1997). *Technobrat: Culture in a Cybernetic Classroom*. Harper Collins

Nehru, J. (1989). *The Discovery of India* (Centenary). Oxford.

Omvedt, G. (2004). *Ambedkar*. Penguin.

Parekh, B. (1991). Nehru and the National Philosophy of India. *Economic and Political Weekly*, 26(1/2), 35–48.

Philip, R. (2016). Revisiting Assessment in the work of Alfred Binet. *Educational Quest. 7(2)*, pp. 79–85.

Poulson, L. (1996). Accountability: A Key-Word in the Discourse of Educational Reform. *Journal of Education Policy*, 11(5), 579–592.

Rudnikski, R. (2000). National/Provincial Gifted Education Policies: Present State, Future Possibilities. In Heller, K.A., Mönks, F.J., Subotnik, R. and Sternberg, R.J. (Ed.), *International Handbook of Giftedness and Talent* (pp. 673–679). Elsevier Science Ltd.

Scharfe, H. (2002). *Education in Ancient India*. Brill.

Schneider, A. L., & Ingram, H. (2008). Social Constructions in the Study of Public Policy. In *Handbook of Constructionist Research* (pp. 189–211). Guilford Press.

Shrimali, K. (2011). *Knowledge Transmission: Processes, Contents and Apparatus in Early India. Social Scientist*, 39(5/6) (May-June 2011), 3–22.

Shulman, D. (2008). Illumination, Imagination, Creativity: Rajasekhara, Kuntaka and Jagannatha on Pratibha. *Journal of Indian Philosophy*, 36(4), 481–505.

Tansley, C. (2011). What Do We Mean by the term "talent" in Talent Management? *Industrial and Commerical Training*, 43(5), 266–274.

Taylor, C. (2009). *The Politics of Recognition. In Multiculturalism.* Princeton University Press.

Terzian, S. (2013). *Science Education and Citizenship: Fairs, Clubs and Talent Searches for American Youth, 1918–1958.* Palgrave Macmillan.

Thorat, S., & Kumar, N. (2008). Introduction. In *B.R. Ambedkar: Perspectives on Social Exclusion and Inclusive Policies* (pp. 1–57). Oxford University Press.

Troyna, B. (1994). Critical Social Research and Education Policy. *British Journal of Educational Studies*, 42(1), 70–83.

Verderey, K. (1993). Whither "Nation" and "Nationalism"? *Daedalus, 122*(2), 37–46.

Williams, R. (2015). *Keywords: A Vocabulary of Culture and Society* (Revised). Oxford University Press.

Zachariah, B. (2001). Uses of Scientific Argument: The Case of "Development" in India. *Economic and Political Weekly*, 36(39), 3689–3702.

# 3 In Search of 'Talent'
## Constructions in Indian Education Policy

*'It is universally accepted that children with special talent or aptitude should be provided opportunities to proceed at a faster pace, by making good quality education available to them, irrespective of their capacity to pay for it'.*
(Government of India, 1986, p. 16)

This Austenesque statement occurs in the second National Education Policy of India. Nonetheless, the identification and support of talented students in the Indian context have not always been 'a truth that has been universally acknowledged'. Indeed, the very sparsity of the discourse on talent in policy literature seems to speak to the opposite. For instance, India has not had a policy document solely targeted at the population of the 'talented' (unlike, e.g., the Marland Report in the United States in 1972, which was the first of many policy papers on 'gifted' children). Despite this, one may nonetheless discern patterns in the ways in which 'talent' has been spoken of and known.

When we think of policy, what comes to mind are 'governing texts' which get circulated within 'a social field' and 'bind people to their mandates' (Levinson et al., 2009, p. 767). Policies tend to represent 'the more superficial, outward expressions of the nation-state, perhaps the most visible narratives of state activity and identity' (Stein, 2004, p. 5). Situating 'talent' within the discursive context of policy literature is a considerably complex endeavour. We have had three national policies of education (1968, 1986, and 2020) in post-Independence India. However, the scale of Indian education policy is far more than just these three texts, in so far as it encompasses systems of thought and action, which have the power to regulate and organize behaviour (Levinson et al., 2009). Apart from the national policies, we can also consider the Constitution of India, the five-year plans, various commission reports, review committee reports, the National Curriculum Frameworks (NCFs), reports of government think tanks like the National Knowledge Commission, NITI Aayog (the former Planning Commission), etc. as part of state education policy because 'they construct ways of seeing those affecting or affected by a problem' (Stein, 2004, p. 5).

Policies are 'the results of compromises at various stages (at points of initial influence, in the micropolitics of legislative formulation, in the

DOI: 10.4324/9781003344902-3

parliamentary process and in the politics and micropolitics of interest group articulation)' (Ball, 1993, p. 45). At the same time, politics constantly influences the meanings of policies, especially when key interpreters of the policy change (ministers of state, secretaries, chairpersons of committees, etc.). Also, the problems which the state faces change over time, giving different policies different rates of momentum in effecting change (Ball, 1993).

Having acknowledged this complexity, this chapter will explore policies at their textual level and analyse the discourse on talent by examining two aspects of its representation. The first is what we may call an explicit discourse of talent, i.e., the content of what specific policies recommend or advocate with respect to one particular category, i.e. talented students. The implicit discourse lies in the representations of talent across policies. It lies in the ways in which various concepts and contexts are integrated using language to create a 'particular sort of socially recognizable identity' (Gee, 1999, p. 29). In other words, this level of discourse analysis explores the linguistic morphology associated with 'talent' in a policy text.

This chapter is structured as follows. After a brief note on the method of content analysis, the explicit discourse of talent is first presented through its evolution and features in 14 landmark policy documents post-Independence. The second part of the chapter is a deeper linguistic engagement with the content of two policy documents, i.e. the Kothari Commission Reports (1966) and the Report to the Nation (2006–2009) of the National Knowledge Commission, and undertakes a close examination of the use of verbs, adjectives, nouns, and synonyms used in relation to the 'talent' in both. Through identifying some patterns in the representation of talent in these two discursive modes, this chapter hopes to draw some inferences about the construct of the talented student in Indian policy.

## A Note on Method

Eleven landmark policy documents in Indian education after 1947 were selected for content analysis of the patterns of word usages, especially those related to 'talent'. Using the software, QSR-NVivo 10, the first exercise that was carried out was a search of the words 'talent', 'talents', and 'talented' in the selected policy documents. Taken together, there were 308 references to talent, which was less than 0.01% of the total word count of these documents. Each of these references was documented and analysed with respect to the content of the whole policy. The documents were then classified according to the maximum number of references to these words.

Among the 11 policy documents (from the Ministry of Education, or the Ministry of Human Resource Development (MHRD), as it was previously called) which were analysed for their explicit discourse on talent, the report of the Education Commission (1966), i.e. the Kothari Commission Report, contained the maximum number of references to the word 'talent'. With 127 mentions, this comprises 43% of the total references to talent among these documents. The extent to which this report stands out may be seen in the fact that the second-highest number of references to 'talent' occur in the Acharya Ramamurti Committee report at 39 mentions, i.e. 13.2%.

Rather than compare the discourse on 'talent' within two MHRD policy documents, it was decided to examine a text from a different source. The main reason for this choice was to examine if there are commonalities in the representation of 'talent' across policy documents emerging from different sources. Between 2006 and 2009, the National Knowledge Commission, the independent think tank instituted by the former Prime Minister, Manmohan Singh, prepared several reports on the state of higher education in the country. The compilation of these reports, called the Report to the Nation (2006–2009), was selected because of a significant number of references to the word 'talent', i.e. 62 mentions. A second reason for the choice of these documents is that the Kothari Commission Reports (hereby abbreviated to ECR – Education Commission Reports) and the National Knowledge Commission Reports (abbreviated to NKCR) are separated by more than 40 years. They also almost bracket the period of the NTSE, which is studied in this research, i.e. 1963–2013. However, a significant difference was that ECR surveys the whole field of education, while the NKCR's focus was on the state of higher education in the country (Table 3.1).

To take the content analysis forward, each instance of the use of the words 'talent', 'talents', and 'talented', which was highlighted by the NVivo software was examined in both sets of reports. The proximate context (a certain number of words used before and after the selected term) in each case was also examined. The main categories which were used for the classification of these proximate words were based on the following parts of speech: adjectives, verbs, nouns, and synonyms. The words falling in each

*Table 3.1* References to talent in select policy documents

| Policy document | References to talent/talents/talented |
| --- | --- |
| Education Commission Report 1964–1966 (1966) | 127 |
| National Knowledge Commission Report to the Nation 2006–2009 (2009) | 62 |
| Acharya Ramamurti Committee (1990) | 39 |
| Programme of Action for NPE 1986 (1992) | 18 |
| National Policy on Education (2020) | 13 |
| Secondary Education Commission | 11 |
| National Curriculum Framework (2005) | 10 |
| National Policy on Education (1986) | 9 |
| National Curriculum Framework (1988) | 6 |
| Yashpal Committee Report (1993) | 5 |
| National Curriculum Framework (2000) | 5 |
| National Policy on Education (1968) | 3 |

category were again sorted into sub-categories based on themes which seemed to emerge in terms of how talent was referenced.

The classification and analysis of synonyms, however, were different from the case of the other parts of speech. When it was observed that there were a number of words which were similar to talent that were used in the reports, a list of synonyms of talent was first identified using a thesaurus. These then were used as keywords for text searches through NVivo in these documents. The number of references to each of these was counted and their sum was treated as representing *the total vocabulary associated with competence*, i.e. the potential for doing, or the execution of, some task both efficiently and successfully. The number of references per term was then converted to a percent value of the total vocabulary, so as to enable comparison between both reports.

## The Explicit Discourse of 'Talent' in Indian Policy Documents

This section examines the tenor of the representation of talent in select policy documents that were critical in shaping educational discourse after Indian independence. I do not summarize the content of these policies because they touch on diverse and numerous topics apart from 'talent'. Rather, the general representation of the 'talented student' in each of these policies is presented here, along with a few select statements which indicate the policy's perspective on the issue. In doing so, I aim to make visible certain trends in the evolution of the discourse of 'talent' in Indian educational policy.

The Report of the Secondary Education Commission (the Mudaliar Commission), 1952–1953 refers to talent consistently in the sense of potential possessed by all learners. There is also an assumption that the fruition of this potential is dependent on the way it is directed during adolescence. For example, consider the following statement:

> In view of the fact that education up to the age of 14 has been made free and compulsory under the Constitution, students with a very wide variety of talents will be seeking education in future. This postulates that our Secondary schools should no longer be "single- track" institutions but should offer a diversity of educational programmes calculated to meet varying aptitudes, interests and talents which come into prominence towards the end of the period of compulsory education.
> (Mudaliar Commission, 1953, pp. 26–27)

The other nine references to 'talents' in the document also reiterate that secondary education must be conceptualized so that every student's inner potential might be developed in the best possible way.

A decade later, the Kothari Commission Reports, titled 'Education and National Development' (1966), envision the state's role in its ability to use education 'deliberately' to 'develop more and more potential talent and to harness it to the solution of national problems', thereby eventually creating a 'social and cultural revolution' (1966, p. 7). In this context, the following extract from the report is noteworthy.

> [P]roviding secondary and higher education to all the potentially able students generally sets up a very high target which even affluent societies find it difficult to achieve. It will be obviously beyond our reach, at least in the immediate future, in view of the limited resources available.... In the transitional period, immediate effect should be given to one important implication of this policy, viz., to ensure that all gifted students (at least the top 5 to 15 per cent of all students), who complete primary or secondary education are enabled to study further in institutions of secondary (or higher) education.
>
> (Kothari Commission, 1966, p. 152)

The fundamental difference from the Secondary Education Commission report is its use of the word 'talent' as a way to express a special potential residing in some learners as opposed to others. Therefore 'talent' is envisioned as the rationale behind the rationing of existing educational opportunities. There is also an explicit confidence in the 'talented' as the vanguard whose special aptitudes would hasten the processes of modernization in the country. There is also a confidence in the design of suitable testing instruments which enable the identification of the 'talented'. Additionally, the discourse of 'talent' is also marshalled in the kind of teachers and personnel whom the education system hopes to attract to itself.

The 'identification of talent' is one of the 17 points identified on the basis of the Education Commission Reports in the first National Policy on Education (NPE) (1968) as the basis for the development of education by the Government. The policy acknowledges and endorses the second view of 'talent', as evident in statements like the following:

> For the cultivation of excellence, it is necessary that talent in diverse fields should be identified at as early an age as possible, and every stimulus and opportunity given for its full development.
>
> (Government of India, 1968)

Not only was such a view endorsed, J. P. Naik, who was the Member-Secretary of the Commission, writes how it was one of the few points on which there was no disagreement in the politically charged situation in which the policy was developed (Naik, 1982).

The next significant policy document, the white paper 'Challenge of Education: A Policy Perspective', which was circulated by the Ministry of Education in 1985, prior to the development of a revised National Policy of Education, does not use the word 'talent'. However, using the word 'gifted' to refer to the same population, it links learning capabilities, the quality of institutions, and elitism, within a discourse of citizenship and its expectations. In fact, this is the first policy document to interrogate the assumption that an investment in 'talented students' translates into an unqualified benefit for the nation.

> Even amongst the gifted, with their sharper perceptions, who get the opportunity of studying in the IITs or IIMs or the Medical Colleges, at very little cost to themselves, there is no evidence of the expected commitment to social responsibility. The same applies to the products of better-quality schools. In fact, this problem at the school level is even more acute because few elite schools concern themselves with developing a sense of social obligation amongst their pupils... the result is a kind of snobbishness which distances the products of these schools from the realities of their environment.
>
> (Government of India, 1985, p. 9)

Interestingly, the second NPE, which was passed a year later, does not adopt this position (as is clear from the opening statement of this chapter) but rather embraces the need to provide special opportunities for the talented by espousing and further developing the 1968 Policy's perspective on talent. In this context, the 1986 policy recommended the setting up of the residential Navodaya Vidyalayas, whose aim was to achieve

> ... the objective of excellence coupled with equity and social justice (with reservation for the rural areas, SCs and STs), to promote national integration by providing opportunities to talented children from different parts of the country, to live and learn together, to develop their full potential, and, most importantly, to become catalysts of a nation-wide programme of school improvement.

Let us note how the idea of talent (in the 'excellence' discourse) is linked in this text with the nation's ideals of equality and democracy using the means of 'reservation'. The residential Navodaya school is not presented as an island of excellence but rather as a potential microcosm of the Indian 'nation'.

Prepared two years later, the NCF of 1988 does not add anything new to the discourse of talent. However, it implicitly upholds the bell curve model of 'talent' in the context of learning. While it acknowledges the general principle of the 'importance of paying individual attention to each learner and allowing him or her to proceed at a pace suiting to his/her abilities and aptitudes', the recommendations for learners at both ends of the bell curve are not very detailed. An example is given below.

> Pupils with special talents or aptitudes should be provided with opportunities to proceed at a faster pace and good quality education should be provided to them. Similarly, there should be provision for remedial instruction for those who are slow learners.
>
> (NCERT, 1988)

The next significant report for our analysis is the Acharya Ramamurti Committee Report, i.e. the Report of the Committee for the Review of the National Education Policy – 1986. Its report, published in 1990, contains an unusual analysis of the idea of 'talent' because of the detailed manner in which it attempts to demonstrate that 'talent' as a basis for classifying students is a social construction, rather than a natural trait. What passes as 'talent' is in reality the sum of socio-economic and family advantages, which some students possess as opposed to others. The implications of the state supporting the identification and development of the talent of some students are presented as

> a question of equity and social justice... since a majority of rural children grow under the constraint of impoverished conditions and poor schooling, which limit the development of talent, aptitude or merit.
> (Acharya Ramamurti Committee, 1990, p. 90)

Secondly, the report denies the possibility that an entrance test to select children for such a programme would be culture neutral and stated that

> ... entrance tests cannot be accepted as fair tools for identification of special talent or aptitude' in a culturally diverse and stratified society like ours.
> (Acharya Ramamurti Committee, 1990, pp. 90–91)

The report is also unusual in the inclusion of the fact that all the members of the committee did not agree with this analysis of the nurture of talent through the 'Navodaya Vidyalayas'. The dissenters argued that the scope for 'harm' to any ideal of the nation is less, while the potential for good is much larger.

> Any system of residential schools will only cover a small fraction of the total population; nor are one or two Navodaya Vidyalayas in a district likely to deplete the number of talented students in non-residential schools; and no great harm will be done if a small number of village children who do well have the opportunity to go to good residential schools.
> (Acharya Ramamurti Committee, 1990, pp. 92–93)

The next policy paper, the Programme of Action for the National Policy of Education 1986 (1992), returns to a bell curve model and presents a different vision for different students depending on their learning abilities. It unapologetically endorses two standards of 'quality' within the state education system by presenting the different psychological and academic needs of the 'talented'. In its recommendations for Secondary Education, it notes

> There are talented children with pronounced competence in particular fields which may be accompanied by indifference in certain other areas.

Therefore, arrangements for such students cannot be fitted into regular courses of study. Special arrangements for such students will have to provide teaching/learning on a modular basis for every small group of students in a small number of subjects of interest to them. Such arrangements will be characterized by better facilities, higher teacher-student ratio and regular participation by professionals in teaching arrangements.

(Government of India, 1992, p. 44)

The next year saw the release of the report of the Yashpal Committee, *Learning without Burden* (1993). This report contains an explicit critique of the idea of the 'talent search' in the context of the aims of education of all children within a state schooling system. Consider the following example:

With a view to provide incentives to 'high achievers' and 'talented' in different fields, high profile competitions are organised by different departments and institutions in the name of 'talent search', which at the most provide moments of brief glory to the winners but damage the 'ego-strength' of numerous others who participate in the contests at the cost of leisurely pursuit of knowledge at their own pace and in their own ways. The experience of the ignominy of failure on the part of millions of children [has] long term deleterious effect on the personality of the individual and the matrix of society. It would be better to reward group performance so as to convey the message to everyone that excellence in group work rather than individual effort should be the target.

(Yashpal Committee, 1993, p. 18)

The NCF of 2000 does not challenge the bell curve perspective of the previous curriculum framework. With respect to the needs of 'gifted and talented' children, it envisions the role of a curricular programme

[which]...should identify such children, [and which] ... should also nurture their diverse creative abilities by paying special attention. It is also important that the identification and nurturance begins from the earliest stage of education.

(NCERT, 2000, p. 9)

The idea of 'giftedness' and 'talent' which the NCF 2000 espouses is located in a project which merges Indian cultural ideals and Western psychology.

The task of identifying the gifted and the talented must be accomplished on the basis of a broad conceptualization of the process from multiple perspectives, rather than as a search for a unitary human attribute. Not only their IQ (Intelligence Quotient), but also their EQ (Emotional Quotient) and SQ (Spiritual Quotient) ought to be assessed.

(NCERT, 2000, p. 9)

In contrast, the NCF 2005 adopts the perspective that competencies are socially acquired and makes a case for the development of the potential of all students. Taking a position which upholds a democratic vision of learning, it argues as follows:

> Schools… undermine the diverse capabilities and talents of children by categorizing them very early on narrow cognitive criteria. Instead of relating to each child as an individual, early in their lives, children are placed on cognitive berths in the classroom: the 'stars', the average, the below average and the failures. Most often, they never have a chance to get off their berth by themselves. The demonising effect of such labelling is devastating on children.
>
> (NCERT, 2005, p. 86)

## Features of the Explicit Discourse of Talent

One of the immediate observations which emerge in this analysis of the policy documents is that 'talent' is predominantly represented in two ways. The first and predominant perspective is that 'talent' is a special potential for excellence which is present in some students. Six landmark policy documents support this perspective. These include the two National Policies on Education (1968, 1986), two NCFs (1988, 2000), the Education Commission Reports (1966), and the Programme of Action for NPE 1986 (1992). The second idea of talent in the policy literature presents the concept as general competencies and abilities which may be possessed by all learners. In comparison to the first, this is not a prominent perspective and was explicitly articulated only by five policy documents. These include the Report of the Secondary Education Commission (1952–1953), the Acharya Ramamurti Committee Report (1990), Learning without Burden: The Report of the Yashpal Committee (1993), the NCF (2005), and the NPE (2020).

The policies which fall into the first category make the case for developing the potential of talented students from the principle of excellence. Excellence is presented as transcending caste, class, and gender, thereby diminishing the relevance of socio-cultural backgrounds and identities. Excellence is also understood in terms of scale, i.e. a 'top' fraction of the student population. The relationship between this small percentage of the population and the rest is represented as a catalytic one. A catalyst, while being of a small quantity, is a facilitator, which speeds up a chemical reaction. This idea is transposed in the argument for how the development of talented students benefits the larger education system. These students are a vanguard who will quicken the development of excellence in the country. At the same time, the efforts to meet their curricular and other needs by the state 'set the pace' in terms of quality standards for the rest of the educational institutions.

In contrast, the number of policies which refutes the existence of a separate category of students whose potential is worth developing over and above the rest are few in number (i.e. the Acharya Ramamurti Report; the Yashpal Committee Report; the NCF 2005; and to an extent, the National Education Policy 2020, which is analysed at the end of this section). They make this argument from the principles of social equity and justice. These documents manifest great unease with the fact of labelling school children and present this practice as undermining India's commitment to democracy and its translation into the everyday reality of educational institutions. The implication of a differential standard of quality for a few students as opposed to the rest is considered to compromise the aims of education for all the students of the country. Therefore, the issue of scale is addressed as the 'greater good of the greater number'.

Therefore, the explicit discourse on talent posits this validity of the category of the talented student in relation to the project of 'nation-building' with reference to two images of the nation. I have represented the correlation of these two ideas of talent and the nation in an 'ideal typical' fashion. However, in both policy and programmes, these are not as drastically differentiated and mutually exclusive.

The meritocratic and the democratic perspectives of the nation co-exist. One area where we can see the co-existence of this duality is in the perspectives of the state's role in addressing individual disadvantages. For example, even those policies which support the existence of a special category called the 'talented' cannot present a purely meritocratic argument within the Indian context because of the historical and socio-economic disparities which exist within the population. Therefore, they present that the reservation policy is posed as a bridge or a connector between the discourses of meritocracy and that of democracy. Within the context of a historically disadvantaged group, reservations produce a meritocratic method of allowing some members to access opportunities which are denied to the group. Within these policies, the idea of a quota is not presented as antithetical to the principle of meritocracy. In contrast, in the second group of policy documents, which reject the need for special interventions for the talented, reservations are not the primary mode through which the disadvantages which learners bring to the classroom are addressed. Attempts to understand the individual learner's requirements within the constructivist paradigm and the teacher's agency within the classroom are posited within these policy documents as a more substantive way to address inequality.

Perhaps no other policy has been as circumspect in ostensibly appearing to balance the democratic and meritocratic perspective as the National Education Policy (2020). The discourse on 'talent' is characterized by two aspects, i.e. firstly, its representation as something innate to all children and, secondly, its expression as being tied to the interests of the child. In a limited way, this is a continuation of the discursive shift that we see in NCF 2000. Even in the section called 'Support for Gifted Students/Students with special talents', 'talents' are presented as innate qualities present in all children, whose outward manifestation takes the form of '*varying interests, dispositions and capacities*' (Government of India, 2020, p. 19). This coupling with 'interest' occurs six times in the 13 references to talent in the document. The argument essentially is that it is the demonstration of the strength of a student's interests and capacities that entitles her to opportunities 'to pursue that realm beyond the general school curriculum' (Government of India, 2020, p. 19). Teachers are advised to give those students with 'singular talents and/or interests' 'supplementary enrichment material, guidance and encouragement' (Government of India, 2020, p. 19). The coupling of 'interest' and 'talent' tends to place the onus for developing the child's talent on himself or herself and also enables the shift of this discourse to the realm of extra-curricular activities, such as 'topic centered and project based clubs and circles'. (Examples include 'science circles, math circles, music and dance performance circles, chess circles, poetry circles, language circles, drama circles, debate circles, sports circles, eco-clubs, health and well-being clubs/yoga clubs, and so on.) Another related recommendation is for 'high quality residential summer programmes for secondary school students in various subjects'. Admission to these are to be done through what the policy calls 'a rigorous merit-based but equitable' process, so the best students and teachers, including 'from socio-economically disadvantaged groups' can participate.

While the policy does not ostensibly endorse the existence of a separate class of learners who are talented, the relationship between talent, social context, and the process of learning is not explicitly articulated as in the second group of policies. This leads to a marginalization of the relationship between the socio-cultural context of a child and its relationship to how she pursues learning. The population category 'SEDG', i.e. socio-economically disadvantaged group' (which is defined in great detail with reference to various categories of children) acknowledges the existence of structural disadvantages but does not address how these may be reproduced through the curriculum and other teaching–learning processes in the school. In the section on 'Equitable and Inclusive Education', recommendations only fall in the category of infrastructural development and financial assistance to enable 'talented and meritorious students from all SEDGs' (Government of India, 2020, p. 28) to participate in higher education.

## The Implicit Discourse of Talent

The implicit discourse on 'talent' is often hard to perceive because it entrenches ways of seeing people and events as 'real' and without alternatives. For example, 'talent' is usually associated with words such as 'the best', the 'top', the 'young', etc. Words and ideas become 'articulated', i.e. conjoined in certain ways as opposed to others. This is the textual level of the discourse of 'talent' within policy. For example, two policies which hold diametrically opposite views on the existence of 'talented students' might still use the word 'bright students' to mean something good and desirable. The nuances of language therefore allow us to see how the category of 'talent' is constructed in unobtrusive ways within policy texts. This section presents a morphological analysis of the implicit discourse of 'talent', i.e. it identifies, analyses, and describes certain *parts of speech* (verbs, adjectives, collective nouns, and synonyms) as well as patterns of word usage associated with 'talent' in policy documents.

The following section presents a morphological analysis of the implicit discourse of 'talent', i.e. it identifies, analyses, and describes certain *parts of speech* (verbs, adjectives, collective nouns, and synonyms), as well as patterns of word usage associated with 'talent' in these two reports.

## The Use of Verbs

To explore the construction of 'talent', the verbs used in association with talent are a useful entry point. Verbs denote action or some kind of operation/intervention. One may classify the verbs used according to how they build a picture of the talented individual and the kind of action/operation/intervention adopted by the state in response. The verbs identified in the two reports were classified into four categories as visible in Table 3.2. The number of times each verb is mentioned is bracketed. The predominant verb (and its synonyms) used in each category was used to highlight an

*Table 3.2* The use of verbs

| Category | Source | Verbs | Implicit picture of talented individual | Nature of operation |
|---|---|---|---|---|
| Category 1 | ECR | 'To identify' (11), 'To discover' (4), 'To search [for]' (4), 'to spot' (1), 'to locate' (1), 'to catch' (1) | Treasure | Discovery |
| | NKCR | 'To identify' (1) | | |
| Category 2 | ECR | 'to develop' (12), 'to build [up]' (2), 'to enable' (1), 'to provide [for]' (1), 'to promote' (1), 'to organize [for]' (1), 'to cultivate' (1) | Greenhouse plant | Nurture |
| | NKCR | 'to encourage' (1) | | |
| Category 3 | ECR | 'to provide', 'to mobilize', 'to harness', 'to use', 'to starve [of]' (1) | Resource/Source of Energy | Utilization |
| | NKCR | 'to utilize' (2), 'to tap' (1) | | |
| Category 4 | ECR | 'to attract' (9), 'to draw' (3), 'to induce' (2), 'to feed [back]' (1), 'to retain' (1), 'to cast [the net for]' (1) | Magnet | Direction |
| | NKCR | 'to attract' (17), 'to retain' (10)', 'to recruit' (1), 'to guide' (1) | | |

underlying metaphor for the talented individual and the state's response to this construction implied in the policy literature.

Category 1 contains verbs ('to identify' (11), 'to discover' (4), 'to search [for]' (4), 'to spot' (1), 'to locate' (1)) which relate to the 'talented individual' as someone who is of great value but whose potential is simultaneously not immediately visible. The most predominant verb used in this context is 'to identify' an operation which includes the aspects of searching for and recognizing an entity/quality. The 'talented individual' is like a hidden treasure, which must be discovered. Therefore, in the policy context, an implicit narrative in the use of these verbs is the need for some external agency which recognizes the true worth of the individual. The state is presented as fulfilling this role. This kind of representation of the talented person also finds echoes in the popular cultural vocabulary of the Indian context. For example, a colloquial reference to this kind of perspective on talent lies in the Hindi proverb 'gutadi ke lal', which implies the priceless gem hidden in a poor man's quilt. The proverb evokes both the hidden nature of talent in unexpected places and the vulnerability of the talented individual to the vicissitudes which may occur due to adversity. The confidence of the state in its ability to recognize the deserving beneficiaries of the label 'talented' is a very dominant discourse in the ECR, while it is considerably more muted in NKCR.

The second category of verbs ('to develop', 'to build', 'to enable', 'to provide [for]', 'to promote', 'to organize [for]', 'to cultivate', 'to encourage') contextualizes the potential of the talented with respect to the future. While not a very dominant theme in the NKCR, it is an important narrative in the ECR, where it is suggested that the talented students should be considered as 'wards of the state' and given special opportunities for flourishing (though not separate schooling), given the concerns about the poor and often adverse conditions of schooling during the 1960s. For example, consider the recurrence of the verb 'to develop' 12 times in the ECR. The underlying perspective is that of the need to nurture talent through a certain type of greenhouse treatment. The horticultural metaphor is deliberately used in a recurring vocabulary associated with the flourishing and wastage of talent in policy. It is evident in common phrases like 'the fruition of talent', 'the flowering of talent', 'the growth of talent', 'the withering away of talent', and 'the neglect of talent', which pepper policy literature. Therefore, the state is presented as being responsible for providing the requisite care required by the talented individual, just as vulnerable plants require customized attention for their optimum development and their protection from adversity.

The third category of verbs (to provide', 'to mobilize', 'to harness', 'to use', 'to starve [of]', 'to utilize', etc.) highlights the modes of utilizing the potential of the talented individual. The metaphor for talent which operates here is broadly that of a resource or a source of energy. There is an urgency which is attached to its proper utilization for the health of the Indian body politic and institutions. This aspect also draws attention to the power of the state to aggregate dispersed individuals and provide contexts for collectivizing their potential. This assumption is that this aggregation then facilitates the attachment of these individuals to various contexts of national problem solving which will benefit from their abilities. This discourse is present both in the ECR and in the NKCR, though it is much richer in the former. (This point is further developed in the discussion of collective nouns used in the reports).

The fourth category of verbs (to attract', 'to draw', 'to induce', 'to feed [back]', 'to retain', 'to cast the net [for']', to guide, to recruit) draws attention to the response of the talented individuals to external direction. The predominance of the verbs 'to attract', 'to draw', and 'to induce' evokes a language reminiscent of the physical operations conducted on magnetic substances. Therefore, they highlight a particular recognition of the agency of the talented person by the state in the policy literature. The state treats the members of this category as endowed with a very strong sense of what is positive as well as negative for their own personal development.

These verbs occurred in the context of the recruitment of teachers for school education and the choices made by students and faculty with respect to university education. They occur in questions like: 'How does one attract the 'best talent' to science?' Or 'how does one attract the 'top talent' to teaching?' Just as the nature of the context draws or repels materials with magnetic properties, the discourse then is structured around creating the kinds of environments which will not just draw but also continue to retain these individuals. This is a major thematic focus of the NKCR's approach to talent in general and is reflected in the fact that 'to attract' talent is used 17 times, especially in the context of research in mathematics and the sciences as well as in relation to the recruitment of faculty and students to Indian Universities. However, this is also a focus in the ECR, given that 'to attract' (mentioned 9 times) is its third most frequently used verb in relation to 'talent'. However, in the ECR, the narrative of attracting 'talent' in the ECR revolves around preventing the talented 'citizen' from leaving the country and seeking greener pastures elsewhere (i.e. the fear of brain drain). In contrast, the immense shift in the worldview associated with the 'attraction' of talent in the NKCR is its acceptance (and at times, endorsement) of the talented individual as a 'consumer'. The onus is on the state/university to create conditions which are competitive in terms of remuneration and recognition to other opportunities through which individuals may seek to find fulfilment.

## The Use of Adjectives

The role of the adjective is to give more information about the object that it qualifies and therefore, it allows us to further explore the assumptions associated with 'talent' in policy discourse. One can identify 21 different adjectives coupled with 'talent' in the ECR and NKCR (Table 3.3). There are 15 adjectives used to describe talent in the ECR, as opposed to 10 in the NKCR. There are only 3 adjectives which are used in common in both, i.e.

*Table 3.3* Adjectives associated with 'talent'

| Category | Adjective | Source | Theme |
|---|---|---|---|
| I | Native (4), Innate (2), Natural (2) | ECR | Beliefs about the origins of talent |
| II | New (3), Available (3), Potential (3), Young (3) | ECR | Expectations and possibilities for the |
| | Young (1) | NKCR | future |
| III | Best (10), Top (2), Special (2), Specialized (1), High (1) | ECR | Degrees of talent |
| | Best (1), Quality (1), High (1) | NKCR | |
| IV | Writing(3), Research (1), intellectual (1) | ECR | Domain specific channelization of |
| | Science (2), Software (1), Technologically qualified (1) | NKCR | talent |

'best', 'young', and 'high'. We may classify the adjectives which are used under four different categories, each corresponding to a particular theme. The number of times each is mentioned is brackets.

The first category of adjectives (native, innate, natural) pertains to beliefs about the origins of talent. Notably, these are present only in the ECR. The manner in which they are used indicates a belief that talent exists before or irrespective of socialization and educational interventions. Additionally, such talent is readily amenable to capitalization for various purposes, especially by the state. In contrast, the NKCR does not reflect on the origins of the competencies which it seeks to promote in society. In general, it offers suggestions on how to manage or improve existing ones.

The second category of adjectives (new, available, potential, young) allows us to see the discourse of talent as a future-oriented one, full of expectations and possibilities. The ECR highlights the role of an activist state in harnessing and directing the 'new', 'young', 'potential', 'available', and 'latent' talent of the children and youth of the country. This discourse is exclusively present in the ECR, except for one reference to 'young talent' in the NKCR.

The third category (best, top, special, specialized, high, quality) draws our attention to the usage of degrees of classification within the discourse of talent. This need for hierarchy within the discourse points to two ways in which talent is invoked in the policy. On the one hand, talent is addressed both as a general reference to competencies and abilities which may be possessed *by all learners* and as a special and valuable character of promise in certain areas manifested by *a select group of individuals*. The selective depiction of the talented is present both in the ECR and in the NKCR, though the former boasts of more diversity of adjectives. The power of the concepts of the 'best, the 'top', the 'high', etc. lies in their invocation of 'excellence', which, while 'signifying nothing' in themselves, operate within a 'performance-centric technocratic language' common to business, sport, and educational institutions (Botting, 1997). These words in the context of state policy mark a certain institutional desire, which is both aspired to and routinely claimed. At the level of individuals, these terms are experienced through the technology of the examination, which is assumed to be a 'scientific, fair, and culture neutral' instrument which produces the micro classifications necessary to select individuals for further academic and professional opportunities (Deshpande, 2006, p. 2442). The logic of the cut-off point is used to create a boundary which separates the best and the rest. Similarly, the adjectives, 'special', 'specialized', and 'quality' also enhance the distinctiveness of those who are marked as 'talented'.

The fourth category highlights the characterization and channelling of highly competent individuals based on the task requirements of various domains ('writing', 'research', 'intellectual', 'science', software, 'technologically qualified'). Like the market, the state has the power to facilitate the development of certain disciplines over and above the rest by providing incentives. This representation is present in both ECR and NKCR.

## The Use of Collective Nouns

The implicit assumptions of 'talent' evoked in state education policy are brought together in the images used to describe the talented as a collective group (Table 3.4). The kinds of collective nouns (words which refer to the collection of the 'talented as a whole) which are used consolidate the state vision of this policy category. Five such nouns were identified. These include 'pool of talent', 'reservoir of talent', 'stock of talent', 'generations of talent', and 'men of talent'. Only two of these (pool and reservoir) are common to the ECR and NKCR.

The first category ('pool', 'reservoir', and 'stock') represents the most significant collective nouns in terms of the frequency of occurrence in both documents. Among them, the reference to 'pool of talent' occurs five times in the ECR and thrice in the NKCR. The idea of the 'pool of talent' can also be related to the reference to 'stock of talent' in the ECR and that of 'reservoir of talent' in the NKCR because all three point to an available and continuous supply of individuals. This picture is at the heart of the idea of

*Table 3.4* Collective nouns associated with 'talent'

| Category | Collective noun | Source | Significance |
|---|---|---|---|
| 1 | 'Pool of talent' | ECR (3), NKCR (5) | Resource |
|  | 'Stock of talent'* | ECR (1) |  |
|  | 'Reservoir of talent' | NKCR (1) |  |
| 2 | 'Generations of talent' | ECR (1) | Transmission |
|  | 'Stock of talent' * | ECR (1) |  |
| 3 | Men of talent | ECR (1) | Generalization |

Note: The usage of 'stock' can be interpreted in both categories and therefore is repeated.

'resource', whose etymological origin derives from the Latin word *'resurgere'*, which means a spring that continuously rises from the ground. Therefore, resource implies a supply that can be drawn upon again and again, without being exhausted (Shiva, 1992). In the context of talent, the idea of the 'pool', the 'reservoir', and the 'stock' highlights a statist imagination of a group of individuals who can be harnessed at need for various initiatives of nation-building.

The words 'pool' and 'reservoir' also draw our attention to the power of the state to aggregate and concentrate that which was previously scattered. The idea of a national pool of talent or ability is not just verbally articulated in these policy documents, but it is materialized through various recommendations, schemes, provisions, and programmes as a category of individuals who are identified, aggregated, and channelized for designated purposes of the state.

The second category of nouns draws our attention to the idea of transmission, a theme which was also raised under the use of adjectives which addressed beliefs about the origins of 'talent' in the ECR. The phrases 'generations of talent' and 'stock of talent' both raise the image of unbroken genetic transmission in conjuring its origins, rather than social creation.

Finally, the ECR also has one reference to 'men of talent', which is used as a gendered substitute for the entire group of talented individuals. This does reveal certain embedded assumptions about who the talented might be.

## The Use of Synonyms

Another aspect that we need to keep in mind in identifying the discourse of 'talent' is that this word alone does not exhaust the discourse. When one reads any policy on education, one notices a range of terms to describe a group of children who possess a greater competence in learning. A total of 551 and 309 mentions of words related to talent were identified in the ECR and NKCR, respectively. These are classified in Table 3.5. It was possible to group them into 13 categories. There was a greater linguistic diversity in the ECR, with terms that fall into all 13 categories as opposed to the NKCR, where the terms used fall into only 10 categories.

The first aspect of the usage of these terms ('talent', 'skill', 'ability', 'creativity, 'capacity', 'gift', 'merit', 'intelligence', 'aptitude', 'bright', 'high achieving', 'brilliant', and 'genius') in the reports is that none of them are 'fixed' with a definition. However, they individually and collectively benchmark the good, the desirable, and the positive aspects of competence in an easily recognizable manner. Additionally, their synonymy references and implies one another in a discourse of competence which is very recognizable in everyday life. For example, it is often assumed that a 'talented' person *possesses* certain 'gifts' or 'skills' or 'talents', which are *manifested* in their *readiness to respond* to people, events, and situations. This readiness is understood as 'ability', 'capacity', 'intelligence', or 'aptitude'. The *quality* of their response is treated as being separate from and elevated above the

*Table 3.5* Synonyms for 'talent'

| Expression | References in ECR | References in NKCR |
|---|---|---|
| Talent/talented/talents | 127 | 62 |
| Skill/skills | 89 | 128 |
| Ability/abilities | 80 | 29 |
| Creative/creativity | 85 | 26 |
| Capacity/capacities | 66 | 46 |
| Gifted/gifts | 33 | 2 |
| Merit/meritorious | 30 | 4 |
| Intelligence/intelligence quotient/ intelligent | 16 | 1 |
| Aptitude | 7 | 6 |
| Bright | 6 | 5 |
| Achiever/achievers/"high achieving" | 5 | 0 |
| Brilliant | 5 | 0 |
| Genius | 2 | 0 |

standards of the normal or the ordinary. This extra-ordinariness is signalled by words such as 'brightness', 'brilliance', 'genius', etc. This recognition constructs them as *deserving and entitled* in their claims to socially valued goods and positions, i.e. 'meritorious' or 'achiever'. Therefore, these individual terms take up a position in an ensemble of ideas related to socially desired competence.

Secondly, within the discourse of competence, the choice of a particular word over its synonyms is significant, and the word count analysis reveals that there are patterns in such usages. One can definitely note that the word 'talent' has a significant presence in both reports, separated as they are by 40 years. For example, in the ECR, talent and its variations ('talents'/ 'talented') top the list with 127 mentions. These comprise 23% of the total vocabulary of competence in the ECR. It is followed by 'skill/ skills', which occur 89 times, comprising 16.2% of the expressions related to individual competence. By 2009, talent and its variations come second in usage in the NKCR, where the maximum number of references is to skill/skills, i.e. 128 references or 41.4% of the total number of words associated with competence. However, references to the core idea of 'talent' still comprise 20.1% of the vocabulary of competence (62 mentions in total), which is only a 3% decrease relative to the kind of word space it occupied within this vocabulary in 1966.

Other changes in the pattern of word usage are reflected in the changes associated with the use of the ideas of ability, creativity, and capacity in discussing the competence of individuals. With respect to the order of usage of the terms, 'ability' (14.5%) is the third-most frequently used term in the ECR, followed by 'creativity' (11.8%), and then 'capacity' at 10.9%. The notable difference in the NKCR is that the word 'capacity' comes third in the number of mentions with 46 references, i.e. 14.9% of the total vocabulary of competence. Also notable is the negligible usage of 'gift/giftedness', 'merit/meritorious', and 'intelligence/intelligent/ intelligence quotient' in the NKCR, as compared to their presence in the ECR at 5.9%, 5.4%, and 2.9%, respectively. In both reports, the use of 'bright' and 'aptitude' is insignificant. 'Achiever' and its variations as well as 'brilliant' and 'genius' occur only in the ECR.

## Discussion

I would now like to conclude by raising two important aspects which this study of policy literature has revealed with respect to the evolving idea of the talented student.

### Imagining the Nation

This chapter has drawn attention to the creation of the 'talented student' as a population category deserving of intervention on behalf of the state by post-Independence education policy. In the context of state education in India, which is constrained by severe resource constraints and tremendous inequality between various types of educational institutions, the explicit discourse of 'talent' draws our attention to the question of the kind of 'nation' which one wants to build. I have drawn attention to two predominant interpretations of 'talent' in education policy, i.e. a special potential in some learners versus a generalized potential in all learners. These draw on two contrasting images of the nation in policy. In the first case, talent is linked to an image of the nation as a meritocracy, where the assignation of individuals to positions as well as the distribution of socially valued goods and other rewards is based on one's potential and performance. Such social organization is believed to be the most optimum for the full development of an individual's potential and his or her contribution to the welfare of society. In contrast, in the case of the second meaning of talent, the underlying image of the nation is democratic. Here, every individual's potential for meaningful social contribution is understood in terms beyond his or her current performance. Rather, individuals are contextualized in terms of their histories and social backgrounds, so that factors like caste, class, and gender are understood to be crucial in shaping the ambit of their abilities and opportunities. The explicit discourse of talent therefore sheds one reason why questions of 'talent' tend to be marginal to the concerns mainstream of education policy; i.e., this discourse compels us to examine the meritocratic and democratic imperatives in the project of 'nation-building' and to interrogate whether both are compatible.

### Imagining the Role of the 'Talented' Citizen

The implicit discourse of talent on the other hand draws attention to the conception of the talented student as a citizen, whose potential is available for the use of the nation's interests. The discourse of 'talent' in the implicit policy language of ECR represents the talented individual as extremely valuable and available for the initiatives of nation-building. At the same time, his or her potential is extremely vulnerable to being lost or wasted if not given proper attention. He or she is represented as passive, awaiting the trigger of state intervention for the development of his or her potential. In contrast, by the 2000s, especially in the NKCR, one sees a shift in the depiction of the 'talented'. There is now a greater acknowledgment of the agency of the talented individual as someone with a strong sense of what is desirable for their own personal development. For example, this is reflected in a greater consideration of what kinds of environment 'draw', 'attract', and 'retain' the talented.

One way to understand this change is by tracing the evolution in the discourse of 'knowledge' and its relationship to 'development' over the past 40 years. In the 1960s, post-independence development was understood as economic growth driven by capital-intensive industrialization. This led to a valuation of knowledge and expertise related to science and technology, possessed by scientists and engineers who would lead the processes of agricultural and industrial modernization, comparable to the Euro-American and Soviet Union contexts. In the case of science and mathematics in particular, the popular perception was that success in these endeavours required innate potencies, more than training ('talents', 'aptitudes', 'gifts', 'abilities', etc.). The belief that such potency would manifest itself during childhood and especially adolescence drove the policy of identifying and nurturing such individuals. The assumption was that such nurturance would pay off in terms of their future competence, which would benefit the nation.

In contrast, the idea of the 'knowledge economy' undergirds the perspective of the National Knowledge Commission. This post-industrial discourse argues that 'capitalism has reached a stage where it is no longer manufacturing or industry which will drive economic growth, but rather, service work requiring specialized knowledge such as computer software, digital and communications technology' (Radhakrishnan, 2007). At the heart of this vision is the representation of knowledge as a 'universal development goal' with the concomitant emergence of the new cultural ideal of the 'transnational knowledge professional'. Inherent in this identity is a notion of continuous individual development through 'lifelong learning' (as opposed to 'traditional disciplinary pedagogy') to meet continually evolving demands and challenges. Here, it is not innate potencies but rather socialized competencies which are emphasized. This shift in perspective has also given new life to the vocabulary of 'skills' into which one can be trained and 'capacities' which can be built through the knowledge of human resource management. Many of these ideas also resonate in the National Education Policy, 2020.

Within this changing matrix of what is desirable competence, the resonance of the idea of 'talent' has shifted from the national to the global. The influential definitions of what constitute talent now emerge not just from nation-states but also from multinational companies. For example, the

American Management Consulting firm, McKinsey, released a highly influential report in 1998 where they argued that 'better talent is worth fighting for', one which was subsequently published under the title 'The War for Talent' (Chambers et al., 2001). The term would do justice to the aggressive recruiting strategies of multinational companies in the twenty-first century. What is notable in this context has been the changing response of the Indian state to this phenomenon. Traditionally, the 'brain drain' and the construction of non-resident Indians have been negative, with these individuals being associated with materialism and even 'anti-nationalism'. However, post the 1990s with the opening up of the Indian economy to a liberal regime, the relationship of the 'nation' and non-resident Indians has changed. The perspective of successive central governments has been to determine how the spending power and international connections of these individuals can be harnessed for the economic development of the country. For example, this shift is visible in the kind remarks such as those made by Narendra Modi, in the early years of his Prime Ministership at international venues and forums, where he referred to non-resident Indians (NRIs) as 'brain deposits', rather than a 'brain drain', on the nation (Sirohi, 2015). Therefore, the changes in the implicit discourse of 'talent' at the level of policy reflect a gradual and larger social recognition of the importance of individual identity and choices in shaping the discourse of the nation, rather than the other way around. In other words, one sees a greater recognition that personal potential and entitlement might be experienced individually in ways which might be significantly different from the state's perspective as articulated in official policy and programmatic structures. These differences shape choices and responses to opportunities, a factor which policy must now increasingly take into account.

# References

Acharya Ramamurti Committee. (1990). *Report of the Committee for Review of National Policy on Education 1986 (Acharya Ramamurti Committee Report)*.
Ball, S. (1993). What Is Policy? Texts, Trajectories and Toolboxes. *Discourse, 13*(2), 10–17.
Botting, F. (1997). Culture and Excellence. *Cultural Values, 1*(2), 139–158. https://doi.org/10.1080/14797589709367141
Chambers, E., Handfield-Jones, H., & Axelrod, B. (2001). *The War for Talent*. Harvard Business School Publishing.
Deshpande, S. (2006). Exclusive Inequalities. *Economic and Political Weekly, 41*(24), 2438–2444.
Gee, J. (1999). *Introduction to Discourse Analysis*. Routledge.
Government of India. (1968). *National Policy on Education- 1968*.
Government of India. (1985). *Challenge of Education – A Policy Perspective*.
Government of India. (1986). *National Policy on Education-1986*. New Delhi.
Government of India. (1992). *Programme of Action for the National Policy on Education 1986*.
Government of India. (2020). *National Education Policy 2020*. New Delhi: MHRD.

Kothari Commission. (1966). *Education and National Development: Report of the Education Commission (1964–1966) (Kothari Commission Report)*.

Levinson, B., Sutton, M., & Winstead, T. (2009). Education Policy as a Practice of Power: Theoretical Tools, Ethnographic Methods and Democratic Options. *Education Policy*, 23(6), 767–795.

Ministry of Education. (2020). National Policy on Education- *2020*. New Delhi

Mudaliar Commission. (1953). *Report of Secondary Education Commission of India (Mudaliar Commission Report)*.

Naik, J. P. (1982). *The Education Commission and After*. Allied Publishers.

NCERT. (1988). *National Curriculum Framework 1988*.

NCERT. (2000). *National Curriculum Framework 2000*.

NCERT. (2005). *National Curriculum Framework 2005*.

Radhakrishnan, S. (2007). Rethinking Knowledge for Development: Transnational Knowledge Professionals and the "new" India. *Theory and Society*, 36(2), 141–159. https://doi.org/10.1007/s11186-007-9024-2

Shiva, V. (1992). Resources. In W. Sachs (Ed.), *The Development Dictionary*. Zed books.

Sirohi, S. (2015, September 29). *NRIs Brain Deposit, not Brain Drain: PM Narendra Modi*. The Economic Times.

Stein, S. (2004). *The Culture of Education Policy*. Teacher's College.

Yashpal Committee. (1993). *Learning without Burden*.

# 4   The 'Science' in the Search
## The Legacy of a Brief Experiment

As texts, annual reports are peculiar beasts. They are usually anonymous. In the case of the corporate or the organizational annual report, the absence of the name of the author points to the fact that the organization itself is the author. Therefore, the annual report may be read as a form of autobiography, where the content concerns an account of the self, corporate rather individual, and is integral to its self-fashioning (Davison, 2011, p. 124). Five years into the scheme, the NCERT Annual Report of 1967 strikes a distinctly congratulatory note in its assessment of the evolution of the Science Talent Search. It refers to the first pilot talent search test conducted in 1963 in Delhi as 'an experiment' that 'was successful and was extended to cover the whole of India in 1964'. The terms 'experiment' and 'successful' are revealing. What kind of experiment did the creators of the Science Talent Search believe that they were undertaking?

For an answer, one may turn to five annual reports on the Science Talent Search prepared by the Department of Science Education, National Council of Education Research and Training (NCERT), between 1963 and 1967. These make for a refreshing departure from routine, dry as dust tomes that one might expect from the genre of annual report literature, primarily because these were written as reports of educational research. Each of these contained a forward by the Head of the Department (R. N. Rai till 1965 and M. C. Pant in the reports of 1966 and 1966) and a preface by the author Dr K. N. Saxena (a field advisor with the Department). These paratextual notes frequently articulate the hope that 'teachers, educationists, educational administrators, psychometricians and research workers, interested in Mental Measurement, will find this report interesting and useful' (Saxena, 1965, p. 1).

The report of 1966 addresses the question at the heart of the science talent search, i.e. the 'origin of scientists', in a concise paragraph.

> The perennial question is "Can scientists be identified and nurtured or they are an out-come of pre-designed circumstances and heredity?" The answer to this question is a very complicated one. Many people believe that aptitudes are not inherited and therefore if a suitable environment

DOI: 10.4324/9781003344902-4

is provided, it is possible to nurture this intellectual faculty to a remarkable extent. There is another school of thought which believes that aptitudes are hereditary and are also dependent upon environmental circumstances. It is, however, true that some basic intellectual potentialities should be present in order that an individual can profit by an accelerated environment. Empirical experience has shown that it is possible to spot children and adolescents having superior or above-average mental potentialities and then to nurture these on specific scholastic and vocational lines. If this holds good, it should be possible to pick out individuals having basic mental faculties and nurture their talents according to their specific aptitudes.

(Saxena, 1966, p. 4)

The perspective in this passage reflects a confidence in the amenability of aptitudes to external qualifications as 'basic', 'above average', 'superior', etc. This confidence resonates in all the Science Talent Search reports. These aptitudes are also identifiable through suitable procedures, so that select individuals can be given suitable curricular experiences. All of these, however, hinge on the verb 'spot', whose simplicity hides the paraphernalia and history of professional expertise of educational assessments, including intelligence testing. An important aspect of the approach to 'talent' within the first years of NSTS is how deeply it is shaped by and embedded in a discourse of 'professional expertise'. The official literature repeatedly refers to the importance of modern testing techniques wielded by experts in revealing the presence of scientific aptitude in students.

Often this discourse of the competence of expertise is reinforced by highlighting the limitations of the school teacher. A study conducted by the NCERT's Department of Science Education (DSE) in 1964 indicated that there were very low correlations between the teacher's subjective ratings on personality characteristics (such as self-confidence, initiative, aptitude for science, inventiveness, laboratory skills, and scientific attitude) with the Science Talent Search sub-tests and total score. This discrepancy was attributed to the teachers' lack of an 'independent' and clear idea about the personality traits of individual students due to the large number of students who are put under their charge. Besides, the 'complete indifference' of the teachers regarding the study was also regretfully mentioned (Saxena, 1965).

When the word 'experiment' is used in the institutional discourse of the talent search, it is certainly not used in the sense of a procedure (perhaps novel) that tests a hypothesis, such as the possibility of identifying students with scientific aptitude. A science talent search of this nature was not novel in the Indian context, considering that it was preceded by the Jagdis Bose National Science Talent Search Exam,[1] which had been conducted in West Bengal since 1956. More accurately, 'experiment' and 'successful' can be read as reflections on the capabilities of the then newly established NCERT (September 1, 1961), which conducted the Science Talent Search.

## The Construct of the 'Talented Student'

In 1964, M. C. Chagla, the reputed jurist-diplomat and Minister of Education, described the talent search in the following words, 'We pick talented boys and girls who show an aptitude for science and give them the finest scientific education possible' (quoted in the Proceedings of the Third Meeting of the NCERT, 1964). The 'finest scientific education possible' was indeed a lofty label to attach to the nurturance program created for the selected scholars. Without reading too much into this phrase, one can admit that 'nurture' was an important emphasis of the program during its initial phase. The design of the examination and the nurturance programme were substantiated within the official literature by a particular construction of the talented student. A notable feature of the way that the talented student is depicted is the emphasis of the vulnerability of this cohort if the State did not intervene.

### At Risk from the Rest of the Student Population

The reports repeatedly classify the student population along a normal distribution curve, with learners being divided into three categories: 'the academically bright', 'the mediocres', and 'the slow learners'. A shared educational context is presented as a zero-sum game between the interests of these three categories, with the representation of the academically bright' as being 'handicapped' or 'sacrificed' at 'the altar of the slow learner or the mediocre' (Saxena, 1964, 1965, 1967). Additionally, one is warned that 'in the typical scholastic situation, [the] higher capabilities of [academically bright children] may dwindle away into the limbo of oblivion while adjusting to the mediocres and slow learners' (Saxena, 1965).

Rampal notes that early post-Independence policy documents like the National Curriculum Framework of 1975 and that of 1988 also retain the division of students into 'high achievers' and 'low achievers'. A feature of these analyses is the notion that the fast need 'enrichment', while the slow need 'remedies'. Such implicit assumptions reinforce deeply discriminatory notions that the slow were sick while the fast deserved 'enrichment' (Rampal, 2009). This seems to be a general feature of policy representations of gifted and talented students even in other countries. For example, the title of the 2015 Templeton report on the status of gifted education in the United States sums up the nature of its argument, i.e. 'A Nation Deceived: How Schools Hold Back America's brightest students'.

There is an emphasis on the lack of capability on the part of the regular school teacher in providing 'high-level instruction in different areas of basic sciences' and 'adequate information about modern research' (Saxena, 1965, p. 17). This was partly attributed to their lack of 'understanding [of] the psychology of gifted children and how to provide challenging opportunities to these students' (Saxena, 1965, p. 17).

### *At Risk in the General Education System*

The responsibility on the State to prevent wastage of abilities is highlighted by using language that suggests that these will stagnate and deteriorate if they are not suitably nurtured. Consider this example.

> It is... an admitted fact that if we do not provide challenging educational and extramural programs to the exceptionally advanced children, their intellectual development and growth will not be commensurate with the extensive faculties they possess.
>
> (Saxena, 1967, p. 17)

In the case of the NSTS, the solution is presented as an accelerated program of science education, where scholars find 'peers' who would motivate and challenge one another, as opposed to the mixed ability grouping in schools and colleges who were presented as 'hindrances' for the 'talented'. The case for a specially designed nurturance program for the talented is also made through references that display a general institutional disillusionment. While this is evident in all the NSTS reports, it manifestly increased between the years 1966 and 1967. While there is distrust of the school teacher's abilities, the reports also recognize that teachers themselves might be constrained by curricular and infrastructural limitations such as poor laboratory, library, and audio-visual facilities, despite their best intentions. For instance, consider how the involvement of teachers is represented in the following example. Students could consult teachers when they prepared their projects for the NSTS. The report of 1965 comments on this practice in this manner:

> The standards achieved in the submission of project reports by the students... has made it clear that there is no dearth of talents in the country and that there are a number of teachers who utilize their leisure hours to motivate talented students in the best possible way to develop their potentialities in the real sense of the term.
>
> (Saxena, 1965, p. 45)

What is interesting about such a comment is the author's assessment that a teacher has no space within the system to provide such encouragement but rather needs to use his or her leisure to do so. Here is another example from the report of 1966, which explicitly analyses some institutional issues.

> The students who get the National Science Talent Search award are ultimately admitted to such institutions where age-old traditional curricula, methods of teaching and evaluation exist. Secondly, there are hardly any opportunities for talented scholars to work independently on themes of

their own choice. The load of memorization is so heavy that it seems impossible for brilliant students to do any original work based on creativity and intellectual sagacity. These are some facts of reality which cannot be overlooked while analyzing the results of this scheme.

(Saxena, 1966, p. 6)

## At Risk if not Identified at the Right Age

The statements in the Science Talent Search reports on the age of intervention help us understand the vision of nurture that underlies the program. Initially, the age at which the test should be conducted was based on the assumption the aptitude of a child for science would be visible by the secondary stage of education. The report of 1964 puts it as follows:

One such vital step is to locate the students with a scientific bent of mind at an early stage and then to nurture these talents so that they may develop into potential scientists. Empirical experience has shown that the secondary stage of education appears to be the right stage for identifying the talents in the basic sciences. It is at the secondary stage of education that the potentialities of the child unfold and reveal themselves and could be helped to grow to the full extent to which the child is capable.

(Saxena, 1964, p. 1)

The NSTS report of 1966 presents a more complex argument for choosing the secondary stage. It compares the merits of selecting students at the secondary stage as opposed to the 'middle stage of school education' (elementary education). Apart from the basis of 'experience', this report also mentions the results of 'experiments' as guiding this decision.

It is a matter of experience that the two important crossroads, when the individuals need maximum educational and vocational guidance, are at the end of the middle stage of school education and at the terminal of the secondary stage of education. Building up our hypotheses on this postulate, it seems necessary that any attempt for providing concentrated guidance and counseling should be primarily pinned down at these two terminal stages. The terminal stage of secondary education seems to be more important because it is at this stage that the individuals have to lay the foundations of a professional career and have to orient their educational studies accordingly. It is an admitted fact of reality that amongst the multitudes of adolescents and youths at this stage, few possess marked mental abilities and a very defined type of aptitude. Experiments have shown that such adequately goal-oriented individuals, if picked-up by suitable techniques, can form the real core of a team of future intelligentsia provided their talents are nurtured comprehensively.

(Saxena, 1966, p. 4)

One noteworthy point that is emphasized here is that there are a fewer number of students who would be likely to manifest 'marked mental abilities' and 'very defined type of aptitude' and are 'adequately goal-oriented' at this stage. These are defined as individual characteristics and not as those which have been shaped by the advantages of education. It is not clear if the author of the report is referring to the phenomenon of the large attrition of students in the Indian context as they progress from the elementary to the secondary level of schooling. But the fact that this aspect is not considered anywhere in the reports provides a glimpse into the bureaucratic rationality which governed the conception of this program. The program was never conceived as an intervention that was embedded in the school system but as one which operated outside of the ambit and, therefore, the limitations of the school and college system. The question of the factors, such as the relationship between the age of testing and the available pool of talent which can be tested, was therefore not initially addressed.

### In Need of an Accelerated Program of Science Education

In contrast, the need for accelerated science education and the way that the NSTS meets it is presented using a model of nurturing scientific talent through mentorship by practitioners in the field, especially scientists. Access to the close-knit community of scientists, in general, is regulated strictly through age, college degrees, entrance tests, interviews, peer sanctions, geography, etc. But *'acceleration'* in the nurturance program of the NSTS implied the faster induction of the scholars into this community. Part of the reason was also that its prominent gatekeepers were also actively involved in shaping this scheme. This included scientists such as D.S. Kothari. Interestingly, though the scheme is currently no longer restricted to *'science talent'*, 'acceleration' in this sense is still available for winners of the National Talent Search Examination (NTSE) who choose to pursue science as a career. They are automatically eligible for many schemes of the Department of Science and Technology (such as the Inspired Science Pursuit for Innovative Research (INSPIRE) program), which permit admissions to certain elite science research institutes and so on.

Considering that the idea of the academically talented student is posited against the 'mediocre' and 'slow' learners, 'acceleration' also implicitly points to a group of 'peers' who do not slow down their development. However, this idea is treated with considerable caution in the reports of 1964 and 1965. In exploring various options of nurturing the talented, the idea of separate schools and colleges for the gifted is rejected on the basis that these may help 'the academically talented students to emerge out as a separate social group', 'which is a genuine danger' 'for a democratic country like India'. One almost senses the kind of disquiet about such an idea as that which spurred Michael Young's dystopic vision in the *Rise of the Meritocracy*. The summer school is presented as a happy alternative in this regard – enough to motivate the scholars about the existence of more of their tribe but not long enough for any significant social danger.

This representation of the danger represented by the 'talented' is a subterranean theme throughout the reports. While there is an ostensible effusiveness about their abilities, there is a repeated mention of how there is a potential for their energies to be sidetracked into 'unfruitful and socially unacceptable channels'. The most stark example of this fear is embodied in the representation of some winners as 'highly emotional idealists'. On being encouraged to do some good work, they completely absorb themselves in that work and forget the hard facts of life which vitally affect them and perhaps do some creative work in those topics. They easily forget the fact that they should study for the [regular college] examination too (whether it is good or bad at present), a fact which is glaringly brought to their notice every day by the teaching...in the classroom.... If we look at the life of an adult too, we find that a sentimental and idealist behaviour to the extent of neglect of routine duties in life makes a slightly unbalanced personality. (Saxena, 1967, pp. 253–254).

Perhaps to reassure themselves, the test designers ruminated whether science talent search winners really become successful scientists, and they turned to the example of the Westinghouse Science Talent Search, which was more than 20 years old at the time of the inception of the NTSE. Statistics from the latter were used to prove that successful winners could be stable citizens who were models of academic and marital productivity. They evidently contributed to the advancement of their disciplines, were married to spouses who have comparable educational backgrounds, and had four or five children. Women who won the Westinghouse Talent Search awards were depicted as having worked before marriage and as not lost to science even if they temporarily retired to care for their children 'because they [kept track of] their science themselves and through their husbands' work' (Saxena, 1964).

## The Design of the Examination

From the outset, the NSTS reports made much of the fact that the mode of testing in the talent search was distinct from the school examinations. Consider the following example:

> The school marks measure primarily the factual achievement (with imperfect reliability) whereas the STS tests are intended to gauge scientific aptitude (with greater reliability) rather than unpredictable and oscillating scholastic achievement.
>
> (Saxena, 1965, p. 52)

The distinction of the NSTS, which lay in the reliability of the examination, was represented in its embrace of 'modern testing techniques' which were supposed to reveal the presence of talent in the students. For example, the very first report of the NSTS from 1963 notes the high value of the reliability coefficient of the Science Aptitude Test (SAT) in rather purple prose, as is evident below.

> it is natural that in such perfect tools of measuring scientific aptitude, this high-reliability coefficient (which is a very good measure of the internal consistency of individual items also) is expected.
>
> (Saxena, 1963, p. 11)

Between 1964 and 1976, the Science Talent Search Examination was comprised of three parts: a project to be submitted, the written examination comprising of the SAT, and an essay. These were to be followed by an interview for those who cleared the written examination.

The project could be undertaken by a student and submitted in the form of a report along with one's application. The project could be based on experiments carried out by the students or the systematic observation, collection, and interpretation of data related to any particular phenomenon. They were given the option of consulting their teachers. Students could also work in pairs or larger teams, but each individual had to submit their project report. From the organizers' perspective, the purpose of the project was the demonstration of the 'originality of the ideas of the examinee together with his creative experimental attitude' (Saxena, 1966, p. 40). A good project report was considered 'a vital tool of selection for singling out potential students as distinguished from mediocres' (Saxena, 1967, p. 42).

The written examination consisted of two parts. The first part was a SAT comprising 160 multiple-choice questions (MCQs), drawn from various areas including 'physics, chemistry, mathematics, botany, zoology, history, and philosophy of science, biochemistry, biophysics, physiology and hygiene, astronomy, geology, meteorology, agriculture, and engineering' (Raina, 1991, p. 33). The questions were intended to examine the 'powers of comprehension, reasoning, critical thinking, analysis and synthesis, rather than factual knowledge alone' (Raina, 1991, p. 33).

In the early 1960s, this multiple-choice format itself was a novelty in the Indian examination scenario. The prominent pattern of scholastic examinations was the essay type, demanding that students recall and reproduce substantial chunks of information. However, during the 1940s, standardized testing had itself gained legitimacy as well as visibility as a viable means of classification. This was primarily due to the US Army, which had commissioned special intelligence tests to sort through, recruit, and assign its vast number of candidates during the Second World War (Carson, 2007).

The decision to include a multiple-choice test as part of the NSTS was not one without controversy. Much of the five reports of the NSTS from 1963 to 1967 attempts to convince the reader of the superiority of the multiple-choice format over the essay type of questions in 'empirically [judging]… the ability and interest in a particular branch of science' (Saxena, 1966, p. 30). The merit of these items over those of 'recall, completion and matching types' was presented in their ability to evoke 'the powers of critical thinking in students…. [thereby providing] better challenging situations' (Saxena, 1965, p. 6). It was acknowledged that this format of multiple-choice introduced the 'danger of guessing' (Saxena, 1965, p. 6), which would defeat the purpose of trying to identify the critical thinkers from the rest.

However, the designers of the test were optimistic that guessing could be reduced by proper instructions and by using a statistical formula to incorporate this aspect. Nevertheless, Saxena writes in the report of 1966 that there were many associated with the test who belonged to 'the school of thought [which held] that the powers of comprehension, organization of thoughts, and above all, the ability to express the thoughts in words' (Saxena, 1966, p. 9) were important aspects in a talent search. Finally, as a compromise, the examination included a second part, i.e. the essay component. The essay could be written in English or any of the regional languages approved by the Constitution. Interestingly, an internal analysis of the 1965 essays demonstrated that only 28.2% wrote it in English, whereas 46.11% of students answered in Hindi. The rest of the scholars answered in other regional languages (Bengali, Tamil, Telegu, Marathi, Kannada, Urdu, and Malayalam in decreasing order) (Saxena, 1966). From 1969 onwards, the main exam was held in several regional languages.

Those who scored above a certain cut-off in the written exam were called for an interview. An expert committee including scientists and university science professors interrogated candidates on 'various current topics, especially related to everyday life and the immediate environment of the students' (Saxena, 1963, p. 74). A primary objective was to discover the practical problem-solving abilities of candidates. They were even encouraged to answer those questions in writing which they could not solve during the interview period. The interview assessed students on personality traits such as 'self-confidence, verbal confidence, and fluency, general intelligence, depth of interest in science, problem-solving ability, and the ability to think and comprehend the problem clearly' (Saxena, 1963, p. 5). The reports mention that candidates had the freedom to answer in any language of their choice, whether English or their regional language.

The assertion regarding the uniqueness of the NSTS examination strategy is also sketched through the novelty of the content of the examination (its thought-type questions and its search for extra-curricular knowledge) and the expertise of its designers. These are explored in some detail below.

## The Use of 'Thought-Type' Questions

The SAT is frequently described as being composed of 'factual' and 'thought' type questions (from the 1963 report onwards). Consider an exemplary statement:

...[The] thought type items of the [Science Aptitude Test] have proved to be more discriminating than the factual type, which indicates that powers of critical thinking, analysis-synthesis, and reasoning occupy an important place in the effective teaching and learning of science at the secondary stage.

(Saxena, 1964, p. 37)

This statement indicates that 'thought' is more than just general cognitive activity. Rather within the worldview of the NSTS reports, 'thought' is a way of talking about 'scientific reasoning', 'scientific method', and a disposition to 'critical thinking'. These cognitive processes are different from those at play in 'factual'- or 'fact'-type questions.

Consider the following sentence which occurs in the Science Talent Search report of 1966.

These new type of items... [call] forth higher mental powers as against rote memory. It also indicates that the thought type questions elicit spontaneous motivation, which is vital for proper academic distinction.

(Saxena, 1966, p. 38)

There are three points in this statement that require close attention. Firstly, the reference to 'proper academic distinction' towards the end in the above statement is significant as it represents the signal attraction of the 'thought-type' item for the examiner. It is assumed that a question of this kind can discriminate objectively between the high and the low achievers, unlike 'the factual type of questions which can at best test the rote memorization capacity of the individuals' (Saxena, 1965, p. 46). A second aspect of the assumption that the thought-type questions permit objective discrimination between examinees is visible in a term used in the statement: 'spontaneous motivation'. The spontaneity referred to here is the student's response to the test items, which are novel and therefore, challenging. The greater the readiness for a response on the part of the examinee, the more successful he or she is. This idea underlies the original meaning of 'aptitude' as 'aptness' or 'readiness', a meaning which gradually got lost over the etymological evolution of the word (Snow, 2012). A final aspect that this statement brings to our notice is its generation of a hierarchy of mental powers, where memorization is given low status. This denigration of memorization as a way of mastering content is based on an assumption of its inferiority as compared to critical thinking, which is considered the hallmark of science.

However, there are other statements in these reports which hint that such a perspective of testing is not nuanced enough. For example, they admit that some scientific disciplines like chemistry demand the memorization of content. For example, consider the following:

It will be noted that items on chemistry are mostly non-discriminative. One plausible reason for this can be that the content matter of chemistry mostly consists of rote memory and hence it is really difficult for the bright and underachievers to be distinguished on higher mental powers.

(Saxena, 1963, p. 10)

This suggests the awareness that all areas of science need not equally be amenable to the same type of cognitive processes or even the same kinds of testing. However such an insight, while present in the reports, remains marginal to the main discourse.

### Extra-Curricular Knowledge: Interest Versus Strain

The second recurrent theme in the reports of the NSTS between 1963 and 1967 is the test's expectation that the student would possess an extra-curricular awareness and understanding of scientific concepts. The 1963 report states:

Students were expected to have a much broader understanding of scientific topics than is usually required by the routine syllabus.

(Saxena, 1963, p. 3)

Such language, which suggests both expansiveness and immersion, continues in the 1964 report.

Students were expected to have a deeper comprehension of scientific topics than is usually required in the traditional type of examination.

(Saxena, 1964, p. 5)

The reports capture the relationship of the candidate to extra-curricular scientific knowledge through the term 'interest'. The assumption is that an aptitude for science will be manifested in the mastery of 'non-curricular scientific content'. The following statement demonstrates this aspect:

It is distressing to note that majority of the students could not give a proper account of their scientific interest. Only some of them were able to express their interest in the acquisition of scientific knowledge through books and magazines. This indicates that there are very few opportunities in the majority of the institutions and homes for the development of scientific interest of children.

(Saxena, 1963, p. 75)

Here the independent 'acquisition of scientific knowledge' is valorized. Such 'acquisition' is treated as the evidence of 'scientific interest' and is part of the package of traits that are recognized as part of 'science talent'. Notably, this is not treated as something inherent but rather the product of socialization and opportunities.

This becomes more evident if we examine the other significant term also used to represent the relationship of the student to extra-curricular information, i.e. 'strain'. Consider this statement:

> On a closer scrutiny of test items, it was found that the number of items in the interdisciplinary and other non-curricular branches of science should be reduced so that it may not put an undue strain on the students with reference to the paucity of time and lack of actual facilities to acquire non-curricular scientific knowledge.
>
> (Saxena, 1965, p. 8)

Unlike 'scientific interest', which represents a combination of inclination and opportunity, and which spurs the acquisition of knowledge over a longer time, 'strain' is the result of the impetus to acquire knowledge for examination. The reference to 'paucity of time' and 'lack of actual facilities' refers to the conditions under which students prepare for examinations. Therefore, the discourse of 'strain' allows us to see that the development of 'scientific interest' is a subtle operation of social advantage. This awareness then complicates the expectation that a student's encounter with a set of novel test items will reveal his or her readiness/aptitude to meet that challenge. The discourse of 'talent' in the NTSE reports between 1963 and 1967 is not innocent of this understanding of the role played by social advantage in shaping the aspects such as 'interest', which are treated as proxies for 'talent'.

However, it seems that this awareness remained marginal to the discourse of testing. From the first pilot test of 1963 onwards, there was a conscious attempt to weigh the question paper towards the direction of extra-curricular content (based on the Higher Secondary and Senior Cambridge syllabi). The assumption was that 'high achievers [will] have a better acquaintance with [outside information] while low achievers do not possess it' (Saxena, 1963, pp 10-11). Five years later, the report of 1967 states that the students in the top 27% were able to answer questions on new areas much better than on traditional curricular branches because of 'wide reading on the part of brilliant students' (Saxena, 1967, p. 39).

This assumption that extra-curricular knowledge was primarily the result of interest, rather than preparation, was misplaced. For example, in the context of Delhi, the Board Examinations were considered very important, and being a merit scholar in them was considered highly prestigious. Therefore, while coaching for examinations as a parallel industry had not taken off by the mid-1960s, the elite schools of the capital had a tradition of selecting and preparing certain candidates. These were also the candidates who would be selected to write the NTSE and their advanced preparation was an advantage.

## The Design of the Nurturance Component

If one was a Science Talent Scholar, two primary benefits were available. The first was the receipt of a scholarship and the second was the chance to attend a specifically designed nurturance program. Students were also given a scholarship package which was attractive in comparison to those which were offered by engineering colleges and other technical institutions as an additional incentive in opting for basic science. Initially, Rs. 100 per month was to be given during all three years of the BSc program. Later, this was hiked up to Rs. 250 per month during the MSc and Rs. 350 per month during the PhD[2]. This was in addition to an annual book grant of Rs. 100. Nevertheless, the scholarship was not presented as the more important aspect.

In this case, the aim was to provide an experience of an advanced level of instruction, as compared to the routine undergraduate college experience through summer schools at institutions that were renowned for their research programs. The winners of the pilot Science Talent Search attended a workshop with some professors of Delhi University in 1963. However, a more concrete design for a special nurturance program for the scholars emerged by 1964. A seven-day residential program was organized at each of the five centres across the country, i.e. Delhi, Lucknow, Calcutta, Ahmedabad, and Hyderabad. The program primarily consisted of lectures, film shows, and individual and group discussions on science. Scientists and professors from top universities spent time with scholars and discussed contemporary scientific developments. In 1965, this was expanded to a more ambitious program consisting of a month-long residential summer school for the scholars, during each year of their BSc. These schools were held at select prestigious universities or colleges across the country with a professor as Director. The participants attended lectures on modern developments in various branches of science, learned the use of advanced research methods as well as technology, conducted lab experiments, undertook workshop practice, went on field trips to places of scientific interest, and watched special film shows. The summer schools were organized based on disciplines such as physics, chemistry, biology, etc. All the BSc students of a region who had opted for a particular discipline could attend the pertinent school. At the Master's level, scholars had the opportunity to work with professors at national labs or well-known research departments of universities. They could choose which lab or centre they wished to attend. In 1965, five summer schools of one-month duration were held in Delhi, Bhubaneshwar, Meerut, Hyderabad, and Pune. By 1969, their number had swelled to 19 summer schools across the country.

## The Move Away from a Science Talent Search Model

A decade into the scheme, doubts began to emerge if the scheme was meeting its purpose of drawing students into the basic sciences. This section traces the evolution of the NSTS into a general talent search program.

A major driver of this change was the fact that participation in the scheme remained dismal in the 1960s. When the NSTS was first extended throughout the country in 1964, it was expected that around 30,000 students would participate. However, the numbers fell far short of expectation at 7000. For a little more than a decade till 1976, the number of students participating in the examination remained between 7000 and 9000. However, in 1966, it had dropped to as low as 4065 students. Secondly, the bulk of the winners were concentrated in certain pockets of the country, such as Delhi, West Bengal, and Maharashtra, suggesting that the awareness of the scheme had not percolated uniformly across the country.

The attrition rates of scholars from the program were also a matter of concern. An internal study conducted by the NCERT in 1976 noted that 'though declared talented and selected for the talent search award, almost 48% of the awardees go without availing of the scholarship, the main reason being that they do not [take up] Basic Sciences' (NCERT, 1980, p. 146). Instead selected scholars preferred to pursue engineering or medicine instead of the basic sciences. The 1960s was a period of growing disillusionment among professionals of the middle classes regarding opportunities for employment in high-paying and commensurately prestigious institutions. The attractiveness of science as a career opportunity was also suspect because there were not enough spaces in public industries and private companies to accommodate scientists. Coupled with this, there was a growing trend of migration to countries like the United States, Canada, Britain, and Australia in the 1960s due to the transformation of immigration norms due to specific skill-based shortages (Oommen, 1989).

Right from the beginning, the administrators wrestled with the dilemma of how the test was related to the routine examinations and assessments that were carried out by schools. On the one hand, the designers saw the NSTS as a unique test, modern and scientifically validated, and as a better instrument than school examinations. Yet, the talent search interfaced with conventional examinations in two key points. Firstly, no student who had scored less than 55% in science subjects in the previous year could appear for the examination. Secondly, once a scholar was inducted into this program, he or she had to get a first class in the university examinations during the BSc program.

A second reason was the marks that scholars obtained during their undergraduate years. Even though the talent search was designed in a way which was supposed to be fundamentally different from the existing school and university testing styles, a condition for the continued access to the scholarship was that scholars should obtain a first class in the university examinations. As early as 1967, 47% of those who took up the scholarship at the BSc level had dropped out. The Science Talent Search Report of that year radiates unease over the situation.

> Because the brightest students are selected all over the country by the examination, one will naturally expect that a much higher percentage of these should obtain a first class at the university examinations, although the two examinations are of quite different nature.
>
> (Saxena, 1967, p. 250)

Therefore, the picture that emerged was that 'most of the talented candidates leave [the scheme] because of not [opting for] the Basic Sciences; not showing good performance in examinations, taking up jobs and leaving their studies' (NCERT, 1980 p. 147). Ten years after the pilot Science Talent Search exam, in 1973, the NCERT constituted a committee to review the functioning of the scheme. Its primary objective was to consider and modify, if necessary, the selection procedure. The committee recommended that 'highly talented people are a national resource and such people should be encouraged without distinction in the field of activity in which their talent shows itself' (Agarwal & Jain, 2005, p. 32). It was also argued that the country's needs in the areas of technology and medicine were pressing and that the social sciences were also witnessing great advancements. Therefore, in 1977, the scheme was revised and renamed the NTSE. In addition to the basic sciences, selected students could now avail themselves of the scholarship even if they pursued the social sciences and professional courses such as medicine, engineering, and management.

Soon after the program's conversion to the NTSE, the test was no longer conducted by the DSE. After 1984, it was first conducted under the aegis of the Department of Education Measurement, Evaluation and Data Processing (DEMEDP), then the Department of Education Measurement and Evaluation (DEME), and currently under the Education Survey Division (ESD). This change was coupled with a change in the format of testing as well as the design of the nurturance program. Candidates were now required to attempt two multiple-choice written tests, i.e. the Mental Ability Test (MAT), which assessed a student's verbal and logical-reasoning abilities, and the Scholastic Ability Test (SAT), which included questions from science, social science, and mathematics. A language test was also later included. The interview was conducted separately for the science and the social science candidates. After 2004, the interview was also removed as a criterion.

A unique aspect of this transition after 1976 is the absence of any theoretical grounding or reflection on the implications of the shift from a search for 'scientific aptitude' to generalized 'mental ability' and 'scholastic aptitude'. The only time that 'talent' is defined post-1976 in an overt manner (which is in the Annual Report of 2010–2011) states that 'talent refers to the potentiality that manifests itself in a high level of performance in one or more specialized areas' (NCERT, 2011, p. 104). What is notable about the new formulation is the absence of the earlier confidence in how talent will manifest itself. The earlier phase of the scheme was infused with the imagined future of 'the talented student' as a scientist and the efforts needed to nurture this potential, so that benefits could accrue to the nation through his or her contributions.

An effect of this vagueness in conceptualization was that the reference to 'nurture' now no longer includes the commitment to long-term planning of educational experiences and monitoring of the scholar's growth as was in the previous phase. For example, this may be seen in the reduction of the scale of the nurturance program from 19 summer schools of one-month duration conducted across the country and projects undertaken under the supervision of various professors in 1969, to a short residential program ranging from five days to a week's duration length (though there has even been one year where it was as less as a day-long interaction with two professors for just 74 scholars in Delhi). The focus is more on the 'exposure' to 'eminent personalities' and 'content enrichment' of the awardees through lectures and presentations. Those scholars who choose to pursue a graduate degree in the basic sciences have the added advantage of being eligible to attend the special science camps conducted under the Innovation in Science Pursuit for Inspired Research (INSPIRE) program of the Department of Science and Technology (DST).

The scaling down of the nurturance component also had a concomitant erosion of the monetary aspect of the scholarship. The book grant was discontinued. In 1977, the monetary benefits were revised with 200 being given per month for the undergraduate degree and Rs. 300 being given per month for the subsequent Master's and PhD program, along with an annual book grant of Rs. 300. It was as late as 2006–2007 that a long overdue revision of the scholarship grant was undertaken with Rs. 500 per month was allotted for all the students studying in Class IX onwards (irrespective of class/course). This was with an exception for the PhD program, wherein it was paid as per the norms of the University Grants Commission (UGC)[5]. A few years later, the scholarship was further revised, and, currently, the amount is Rs. 1250 per month for Classes XI and XII, with a further increase at the graduate level and above to Rs 2000 per month.

## Discussion

The focus of this chapter was to understand the emergence of the policy category of the 'talented student in science' in the historical context of India during the 1950s and 1960s. Scientists and technologists were essential to a vision of national development driven by agricultural and industrial modernization. The prestige they enjoyed had increased exponentially post the Second World War because the Allied victory was interpreted as the result of their superiority in war machinery and inventions, such as the atom bomb. India's first Prime Minister, Jawaharlal Nehru was personally convinced about the role that science played in national development, and many prominent scientists enjoyed tremendous political influence. The pressing question was how a continuous cadre of scientists could be created and nurtured for the various goals of nation-building. In the case of science and mathematics, in particular, the popular perception was that success in these endeavours required innate potencies, more than training ('talents', 'aptitudes', 'gifts', 'abilities', etc.).

In the literature produced during the first five years of the NSTS, the main programmatic conception of 'scientific potential'/'scientific aptitude'/'scientific talents' was one of an adult quality that exists as potential in school-age children. This potential was understood to be amenable to precise quantification through psychometric testing. A major contribution of the examination to the discourse of testing in India was its justification of the MCQ format as an objective, scientific and efficient replacement for older essay-based questions. Concerning the content on which the students were examined, the terms 'thought-type questions', 'extra-curricular knowledge', and 'scientific interest' represented subterranean qualities in the student, which were to be revealed by careful testing.

However, the constraints and limitations within this conception of talent and its nurture in the NSTS were evident in several ways very early on into the scheme's development. For example, the officials associated with the NSTS were aware that both the thought-type question, which tested 'scientific readiness', and factual questions, which tested the mastery of extra-curricular scientific content, were biased towards the kinds of socialization and opportunities which students had received. Yet this insight remained marginal to the testing strategy. Additionally, the scheme's design of its nurturance program was practically conceptualized as a limited intervention in science education. It was meant to assuage the impact of larger systemic issues such as the content of the curriculum, the pedagogy, and the infrastructural inadequacies, which impeded the talented student's progress in science. Because the conception of the scheme was divorced from the realities of schooling and undergraduate education in India, in effect, it did not make any long-term impact on science education at the school and college levels.

The revision of the scheme in 1977 as the NTSE merely removed the restriction that winners had to only specialize in the basic sciences. To attract scholars, the selected students were allowed to pursue a career in engineering, medicine, and the social sciences. Unlike the first phase of the program, there was no theoretical grounding or reflection on the implications of the shift from a search for 'scientific aptitude' to generalized 'mental ability' and 'scholastic aptitude'. Rather the revised NTSE was characterized by an absence of the earlier confidence in how talent will manifest itself. This lack of theoretical engagement with what was a major break within the discourse of 'talent' in the scheme significantly affected its impact. An important implication was the pragmatic conception of the 'talented student' as one who would be labelled so by virtue of merely passing the NTSE without any articulation of how his or her 'potential destiny' would contribute to the project of nation-building. This redefined general perspective on 'talent', which was divorced from a perspective of its development according to the requirements of a specific discipline, was accompanied by a move away from a long-term planning for the 'nurture' of scholars.

These silences and omissions within the official literature were paralleled by the deterioration of the nurturance program from a month-long residential on to a week-long program of exposure to 'eminent' personalities. In addition, the book grant scheme was discontinued and the revisions of the monetary component of the scholarship were rare. The image of the 'talented student' linked to competence in exam giving was also reinforced by the growth of a coaching industry that grew more robust post the mid-1970s. This changed NCERT's focus to modifying aspects of the test design to reduce the 'coaching effect' and to demonstrate the organization's commitment to objectivity and transparency in the process. The subsequent alterations to the scheme focused almost exclusively on streamlining various aspects in the identification process.

## References

Agarwal, M., & Jain, V. K. (2005). Discussion Paper. In M. Agarwal, & V. K. Jain (Eds.), *Seminar Cum Workshop for the Review of National Talent Search Scheme* (pp. 18–50). Department of Education Measurement and Evaluation, NCERT (unpublished mimeograph).

Carson, J. (2007). *The Measure of Merit*. Princeton University Press.

Davison, J. (2011). Paratextual Framing of the Annual Report: Liminal Literary Conventions and Visual Devices. *Critical Perspectives on Accounting*, 22(2), 118–134. https://doi.org/10.1016/j.cpa.2010.06.010

Oommen, T. K. (1989). India: "Brain Drain" or the Migration of Talent? *International Migration*, 27(3), 411–425.

Raina, M. K. (1991). *The Talented Scholars: Accomplishments and Worldview of the Talented*. Delhi: NCERT.

Rampal, A. (2009). Reaffirming the vision of quality and equality in education. In *Education for All: A Mid Decade Assessment*.

NCERT. (1964). *Proceedings of the Third Meeting of the NCERT*. Delhi: NCERT.

NCERT. (1980). *Annual Report of the NCERT 1979-1980*. Delhi: NCERT.

NCERT. (2011). *Annual Report of the NCERT 2010-2011*. Delhi: NCERT.

Saxena, K. N. (1963). *A Report of the Science Talent Search*.

Saxena, K. N. (1964). *A Report of the Science Talent Search*.

Saxena, K. N. (1965). *A Report of the Science Talent Search*.

Saxena, K. N. (1966). *A Report of the Science Talent Search*.

Saxena, K. N. (1967). *A Report of the Science Talent Search 1967*. Delhi: NCERT.

Snow, R. E. (2012). The Concept of Aptitude. In David E. Wiley, & Richard E. Snow (Eds.), *Improving Inquiry in the Social Sciences: A Volume in Honor of Lee J Cronbach* (pp. 249–294). Routledge.

# 5   The 'National' in the Search

## A Social Geography of Talent Identification

From the beginning of the programme, there were two registers through which the idea of the 'national' was articulated. As we saw in the previous chapter, one was futuristic and depended on the trajectory of the talented student, whose academic contributions in science and technology would add to the material rewards and progress that accrued to the nation. At the same time, a contemporary idea of nation-building rested upon the shoulders of National Council of Education Research and Training (NCERT), a state institution, through its bureaucratic implementation of the program drawing on national, democratic, and inclusive values. With the move away from the science talent model, the second vision of nation-building becomes ascendant.

The first section of the chapter explores how concerns about dwindling participation in the science talent search examination and the concentration of winners in a few metropolitan cities led to the structural revision of the examination's format. The exam was decentralized to a two-level model, the first level of which was conducted at the state level and the second at a national level. The talent search's pursuit of the 'national' as a more comprehensive geographic ambit is critically examined through state-level statistics of winners and participants. The second part of the chapter addresses the other prong of NCERT's vision for the talent search, i.e. the ideal of inclusion, through equitable procedures of selection. The trends of performance in the examination as well as the representation of three cohorts (rural candidates, girls, and SC/ST students) in the official literature are analysed. The manner in which the bureaucratic procedures of the NCERT articulates the meanings of the word 'national' as 'representative', 'inclusive', 'eligible', etc. in the trajectories of the talent search is made visible in this process.

### The Pursuit of National Coverage

When the National Science Talent Search (NSTS) was first extended throughout the country in 1964, it was expected that around 30,000 students would participate. However, the numbers fell far short of expectation at 7000. For a little more than a decade till 1976, the number of

DOI: 10.4324/9781003344902-5

students participating in the examination remained between 7000 and 9000. The numbers flew in the face of a core aspect that underlay the conception and implementation of the talent search. At a very simple level, the idea of 'national' represents the maximum coverage of the search, i.e. an 'all India' test.

Attention to the demographic patterns among the winners focused on different issues during the exam's history. As early as 1964, the role of social and economic contexts in shaping the chances of being selected as a talent search scholar was apparent to the organizers. One of the correlations explored by the Department of Science Education was that between Science Talent Scores and the income level of parents. The study, which is included in the report of 1964, concluded that children coming from higher income groups are likely to do better in the Science Science Talent Search tests than those coming from comparatively low-income groups. It is even speculated that one of the plausible reasons might be that those coming from higher income groups can afford to join better institutions and enjoy facilities that are conducive to the growth of talents. Another disparity which was evident to the organizers was the overwhelmingly urban origins of winners as well as their concentration in a few cities, predominantly Delhi. But these insights remained marginal to the conception of the scheme.

In 1975, the Science Talent Search Scheme was reviewed and along with the decision to expand the scheme to the social sciences, a decision was also taken to allow students in Classes X, XI, and XII to also appear for the examination. Of the total 500 scholarships which were now available, 250 were to be given to winners from Class X, 100 from Class XI, and 150 from Class XII. The number of students who appeared for the examination in 1977 expanded dramatically to 28,955 (nearly three times the participation in the previous year) and continued to increase till 1984. Figure 5.1 illustrates

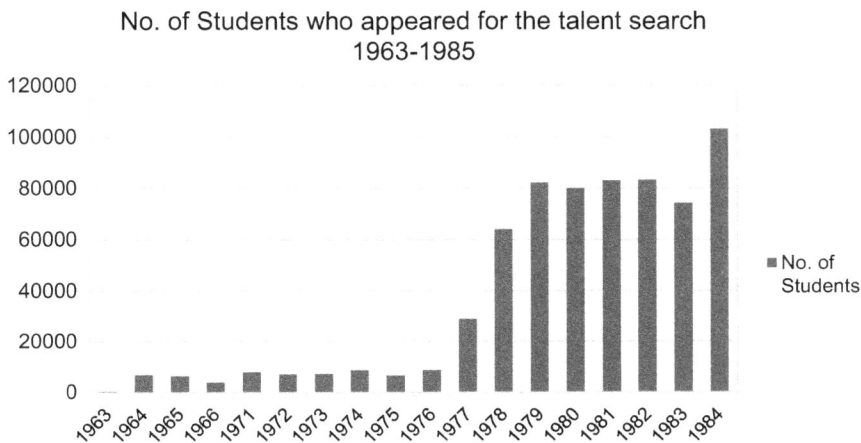

*Figure 5.1* Participation in the talent search, 1964–1984.

the trends with respect to the participation of students in the National Talent Search Examination (NTSE) between 1964 and 1985. Despite some improvement in the number of participations, internal momentum continued to build around the need to address the geographic and social biases within the ambit of the talent search.

However, the next substantive step in the direction of the transformation of the structure of the examination had to wait till the institution of a review committee under the chairmanship of Professor Rais Ahmed, who had been the director of the NCERT and who was at that time the Vice Chairman of the UGC. The aim of the committee had been to review the National Talent Search Scheme in order to make recommendations for its effective implementation from the 7th Five Year Plan onwards. The committee first met in February 1984. By March 1984, there was a focus on the issues of extending the scheme to more disciplines, the syllabi for various subjects, increasing the number of scholarships, reexamining and rationalizing the rate of scholarship, introducing a reservation of seats on a state quota basis, using additional testing devices for selection, and improving the evaluation procedures for selection and nurturing of talent.

After the recommendations of the committee were received, a major outcome was the decision to refashion the talent search examination such that it would be conducted in two stages. The first would be conducted at the state level and a certain quota of students from the first stage would compete at a national-level examination. Scholars would be selected on the basis of the results of this second level. From 1985 onwards, the first level of examinations was conducted at the state level, following which a select number of students appeared for the national-level examination.

## Assessing the Impact of State Quotas in the Talent Search

The Annual Reports of the NCERT from 1985 do not disclose the participation data of students at the first level. Rather only the number of students who appeared at the national level was published. This was an issue that I had raised with the NCERT officials in charge of the test after my initial forays into the annual report literature of the organization. I was told that they did not collect this data but that it remained with the state-level liaison organizations which conducted the first level of the test.

Nevertheless, it was possible to use the data of winners from 1964 to 2005 to build a picture of the state-wise distribution of winners and then to examine if the reorganization of the talent search with state quotas made a substantial difference. Except for the years 1977–1981, 1986, 1989, and 1994, when the reports did not carry the regional spread information, the distribution of 20, 678[1] winners was obtained. Another challenge in depicting the state-wise distribution of winners has been the reorganization of the states in the years after 1964, with several new states being added and this makes it harder to estimate the number of winners from a region.

A major takeaway from this data is the evidence of the concentration of winners in a few states like Maharashtra (3653, i.e. 17.7%), Delhi (2726, i.e. 13%), West Bengal (1723, 8.3%), Uttar Pradesh (1699, i.e. 82%), and Bihar (1402, i.e. 6.78%), which have produced the maximum number of NTS scholars. At the other end, Mizoram (6), Sikkim (9), Jammu and Kashmir (13), Nagaland (21), Arunachal Pradesh (28), and Meghalaya (32) bring up the rear with the minimum number of winners produced during this 40-year period. If we disaggregate this data with respect to the distribution of winners before and after the introduction of the state quota system, we see some shifts.

During the period from 1964 to 1984, the NCERT selected 6667 scholars through one nationwide talent search examination (Figure 5.2). Of these, Delhi produced 31% of all the winners. West Bengal's share was 13.3%, followed by Maharashtra in the third place (9.9%). Subsequently, we have Uttar Pradesh (6.9%), Rajasthan (5.4%), and Bihar (5.3%). At the other end of the spectrum, Nagaland, Arunachal Pradesh, and Sikkim produced no winners between 1964 and 1984. Meghalaya comes next at four winners and then Jammu and Kashmir at five winners.

Figure 5.3 captures the difference in the spread of the 14,011 winners selected between 1985 and 2004. With the allocation of state quotas, Maharashtra (14.5%) settled into the position of the leading producer of winners. Uttar Pradesh (8.9%), Bihar (7.5%), and Karnataka (7%) fell into place behind it. The North East states and Jammu and Kashmir, however, still produced very few winners in comparison. Mizoram (7 winners), Jammu and Kashmir (8 winners), and Sikkim (9 winners) have produced the least number of winners in this second 20-year period. Nagaland did slightly better with 21 winners. Together, these four states have produced only 0.32% of the total number of winners. The North East states, especially Mizoram, Sikkim, and Nagaland have consistently produced a very low number of winners (except Assam).

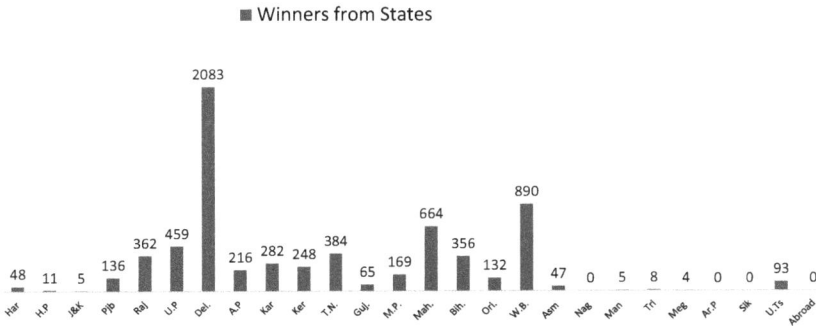

*Figure 5.2* State-wise distribution of winners, 1964–1984.

Source: Compiled from Annual Reports of NCERT (1964–1984)

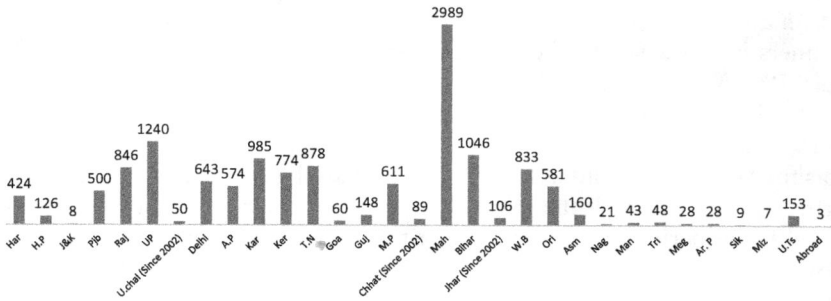

*Figure 5.3* State-wise distribution of winners, 1985–2004.

Source: Compiled from Annual Reports of NCERT (1985–2004)

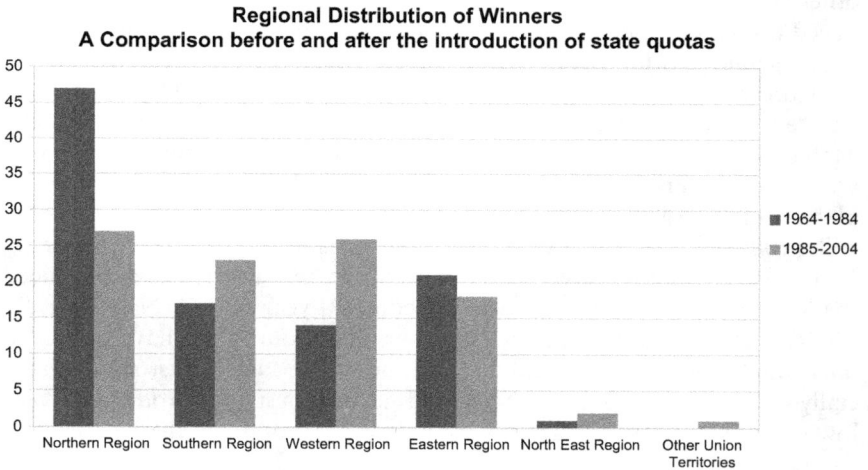

*Figure 5.4* A comparison of the regional distribution of winners before and after 1984.

It is evident that the introduction of the quota system had a significant impact on reducing the level of disparity among the state-wise distribution of winners. If one classifies the various states of India according to regions[1] and then compares the number of winners coming from each region, one can see the more equitable distribution of winners post-1984. Consider the following comparison of winners from 1964 to 1984 and 1985 to 2004 region-wise in Figure 5.4.

But from the Annual Reports of the NCERT itself, we can infer that the state-wise distribution of winners itself is also a misleading figure on its own because winners tended to be clustered in certain schools and certain cities. The Annual report of 1969 specifically picks out 'Delhi', 'Calcutta'

(Kolkata), 'Bombay' (Mumbai), and 'Madras' (Chennai) as metros that produce the most numbers of winners. One reason for the disproportionate representation of winners from Delhi prior to the establishment of state quotas can be attributed to the greater awareness of the examination among schools in Delhi. After all, the scheme did begin as a pilot in Delhi, after which it was expanded nationwide. In the case of West Bengal, the Jagdis Chandra National Science Talent Search Examination had predated the NSTS and therefore, it is plausible that there was a better response to the test from this state.

While Maharashtra has always done well in the exam in terms of the number of winners, its consistent production hints at something more. A former coordinator of the NTSE at the NCERT, when asked about this, mentioned that there is a strong culture and industry around coaching for the NTSE in Maharashtra. One may speculate that the cohort of students who appear for the test and who clear it tend to be from a socio-economic background which can invest in coaching. Since a deep state analysis is not a focus of this chapter, I would just like to flag this data as a possible area for further investigation.

## Hidden Disparities within the State Quota System

A problem with the national data of the distribution of winners is that it obfuscates the nature of distribution of winners within a state. This issue came to my attention as I considered the case of Bihar and Jharkhand (Table 5.1). Between 1990 and 2001, the average number of winners per year from Bihar was 76.5. However, with the creation of Jharkhand and the participation of students from the state since 2002, there is a dramatic drop in the figures of Bihar to an average of 7 students per year for the next three years. Could this data suggest that the number of winners coming from Bihar was actually clustered in a certain number of schools in areas which have now become part of Jharkhand?

The absence of an organized registry of winners prior to 2000 made it hard to explore this aspect for the years preceding this. Since I obtained access to the files after 2000, an examination of the district-wise distribution of NTSE scholars between 2001 and 2005 was undertaken. Working with a dataset of 4811 winners, it was discovered that of the 647 districts

*Table 5.1* Winners from Bihar and Jharkhand

| Year | Bihar | Jharkhand |
|------|-------|-----------|
| 2000 | 135 | – |
| 200 | 105 | – |
| 2002 | 5 | 58 |
| 2003 | 5 | 18 |
| 2004 | 11 | 30 |

of the country (as of 2005), only 352 districts had produced at least one winner during the period under consideration. In others, only 54.4% of the districts of the country had produced at least one winner.

The states of Kerala and Goa are notable in the fact that every district produced a winner. But we also notice that some states stand out in terms of the concentration of a large number of winners (at least 10 to 99 winners per district). This is true, especially of Maharashtra, which shows a relatively high density of winners per district. Other states where the number of winners per district is similarly high are West Bengal and Haryana. Additionally, the extent of the concentration of winners in certain pockets of the country is further reinforced by the fact that 20 districts of the country have produced 40.2% of the total number of winners between 2001 and 2005 (See Table 5.2). The top three districts in terms of producers of winners are Bangalore (Urban), Pune, and Mumbai.

The concentration of winners in certain districts of states was another clue that participation in the examination was highly uneven. In order to understand the patterns in the participation of students in the test, applications under the Right to Information (RTI) were filed to all 35 states and Union Territories to find out the number of students of Class VIII who participated in the NTSE in the years 2009, 2010, and 2011. Data for the participation of students from 9 states was obtained for the year 2009, from 15 states in 2010 and 16 states in 2011.

If we compare the participation of students (see Table 5.3), we see that less than 5% of the total number of students who were enrolled in Class VIII wrote the state-level Talent Search Examination in 13 of 15 states in 2010 and in 14 of 16 states in 2011. Only in the case of Goa do we see a consistent appearance of more than 5% of the eligible students in the state-level test. In 2009, 6.09% of the eligible students participated. In 2010, 5.08% and in 2011, 10.7% of the eligible students wrote the test. Though the data was obtained only for a few states, the patterns of participation

*Table 5.2* Districts producing highest number of winners

| District | State | Winners (2001–2005) |
| --- | --- | --- |
| Bangalore (Urban) | Karnataka | 287 |
| Pune | Maharashtra | 229 |
| Mumbai | Maharashtra | 215 |
| Delhi (9 Districts as of 2005) | Delhi | 197 |
| Chennai | Tamil Nadu | 143 |
| Khurda | Orissa | 142 |
| Kota | Rajasthan | 136 |
| Lucknow | Uttar Pradesh | 129 |
| Durg | Chattisgarh | 126 |
| Jaipur | Rajasthan | 116 |
| Hyderabad | Andhra Pradesh | 110 |
| Faridabad | Haryana | 104 |

*Table 5.3* Participation of eligible students at the state-level talent search

| Percentage of Participants w.r.t. total Class VIII enrolment | 2011 | 2010 |
|---|---|---|
| Below 1% | Rajasthan, Jammu and Kashmir, Jharkhand | Rajasthan, Jammu and Kashmir, Jharkhand |
| 1%–5% | Chhattisgarh, Bihar, Haryana, Karnataka, Himachal Pradesh, Maharashtra, Nagaland, Sikkim, Tamil Nadu, Tripura, Uttarakhand | Chhattisgarh, Bihar, Haryana, Karnataka, Himachal Pradesh, Maharashtra, Sikkim, Tamil Nadu, Tripura, Uttarakhand |
| 5%–10% | Arunachal Pradesh | Nagaland, Goa |
| 10%–15% | Goa | – |

Source: Responses of states to RTI query. The data for Arunachal Pradesh for 2010 was not provided.

suggest that a very small percentage of students who are eligible to write the test in any given year actually appear for the examination.

The disaggregation of winners into districts during the period from 2001 to 2005 and the participation data from states for the years 2010 and 2011 are indicators which challenge the assumption that state-level quotas are more geographically representative and inclusive.

## Demography and the Discourse of 'Inclusion'

Having observed how representation based on geography drove a main alteration in the design of the talent search to a two-level federal and decentralized mode of examination, we now move on to examine the shift from geography to demography in the NTSE's discourse of nation-building.

Beginning from the mid-1970s, there was considerable internal momentum within the NCERT on the need to address social biases with respect to the selection of scholars, in addition to the state-wise concentration of winners. In 1984, a study carried out by M.K. Raina, an NCERT faculty member from the Department of Psychological Foundations, noted that urban, male candidates from socially and economically privileged groups, studying in schools with good infrastructure and English as a medium of instruction were more likely to be successful in the talent search (Raina, 1984). Five years later, Raina (1989) reviewed the scheme again and the results remained similar. His findings demonstrated that scholars belonged to families with relatively high educational and income levels, as reflected in the backgrounds of their fathers who would mostly be engaged in technical, professional, or administrative jobs.

From the 1980s, we can observe a greater focus on the ideas of equity and inclusion. For instance, one sees the reservation of a certain number of scholarships for students from SC and ST categories from 1981. From 2005 to 2010, the purpose statement of the talent search as articulated in the NCERT Annual Reports repeatedly emphasizes the identification of talent from diversified groups like 'rural, urban, boys, girls, socially and economically disadvantaged, children with special needs and school dropouts'. From 2008, 3% of the scholarships are reserved for students who are differently abled (PH candidates). In fact, in 2005, a talent search was also conducted on a one-year trial basis for school dropouts and 30 students were given a scholarship. Most recently, in 2015, the number of scholarships has now been increased to 2000 with the proportion for reservations for SC, ST, and PH students being 15%, 7.5%, and 3%, respectively.

In this section, we trace some trends in the performance of three demographic cohorts, who are constructed as deserving special attention within the program, i.e. the rural students, girls and SC/ST students in the talent search, and analyse their representation within the official literature of the NCERT. In doing so, the aim is to critically engage with the idea of the 'national' that is pursued through espousing democratic aims and ostensibly equitable bureaucratic practices in the second phase of the program as the NTSE.

### Rural Candidates

It is important to note that the analyses of winners along the urban/rural axis have been a consistent demographic reference point in the official literature of the examination. However, there has been no parallel attempt on the part of the officials to examine the geographical distribution of participants. My data received through the RTI applications suggests that there is a substantial under-representation in the participation of rural students in many states.

Table 5.4 classifies the responses which were received from ten states each in 2011, 2012, and 2013, as well as from three Union Territories (2 in 2012 and 1 in 2013). In six out of ten states, less than 50% of participants in the state-level test were from rural backgrounds. However, it is important to note that there are some states which have a good representation of rural participants. Himachal Pradesh saw the highest percentage of rural participants (i.e. between 80% and 90% of the total number of participants) in 2012. Sixty to eighty percent of the participants from Goa and Uttarakhand in these three years have also been from rural backgrounds. Maharashtra had more than 50% of the participants from such backgrounds in 2012 and 2013. Tripura had similar figures for 2011 and 2012. Chhattisgarh had more than 50% rural participants in 2011, as in Karnataka in 2012.

The next indicator that was considered was whether participation in the state-level test was linked to the kind of school that students tended to attend. Table 5.5 presents the responses from 11 states and 3 Union Territories in terms of the percentage of participants who attend government schools. (The data provided by Haryana has been excluded because it clubbed the number of participants attending private aided schools along

*Table 5.4* Percentage of rural participants in the NTSE from various states

| Percentage of participants from rural backgrounds | 2011 | 2012 | 2013 |
|---|---|---|---|
| Below 10% | N/A | Nagaland | Nagaland |
| 10%–20% | N/A | Bihar | Andhra Pradesh |
| 20%–30% | Bihar, Jharkhand | N/A | Bihar |
| 30%–40% | Haryana, Nagaland | Andhra Pradesh, Uttarakhand, *Pondicherry* | N/A |
| 40%–50% | Karnataka, Maharashtra | Haryana, Chhattisgarh | Haryana, Tripura, Chhattisgarh, *D&N Haveli* |
| 50%–60% | Tripura, Chhattisgarh | Maharashtra, Tripura, Karnataka | Maharashtra, Jharkhand |
| 60%–70% | Himachal Pradesh | *Daman and Diu* | Goa, Uttarakhand |
| 70%–80% | Uttarakhand | N/A | N/A |
| 80%–90% | N/A | Himachal Pradesh | N/A |

Sources of Data: Responses to RTI Queries made to states and Statistics of School Education (2009–2010, 2010–2011, 2011–2012), MHRD.

Note: UTs are italicized in the table above.

*Table 5.5* Percentage of participants attending government schools

| State/UT | 2012 | 2013 |
|---|---|---|
| Sikkim | 69.43% | 90.63% |
| Bihar | 86.33% | 71.44% |
| Himachal Pradesh | 73.49% | 50.33% |
| Uttarakhand | 50.17% | 49.75% |
| Karnataka | 47.32% | 47.82% |
| Rajasthan | 28.08% | 41.54% |
| Jharkhand | NA | 40.07% |
| Tripura | 26.98% | 28.61% |
| Maharashtra | 9.94% | 5.84% |
| Andhra Pradesh | 4.68% | 4.89% |
| *Dadra and Nagar Haveli* | 81.69% | 89.81% |
| *Daman and Diu* | 20.00% | 58.99% |
| *Pondicherry* | 45.19% | 48.19% |

Source: RTI responses to query.

with government schools in at least one year). More than 70% of the participants from the states of Sikkim, Bihar, and Himachal Pradesh studied in government schools. Uttarakhand and Karnataka have more than 47% participation of students from government schools. Rajasthan and Jharkhand have more than 40% government school participants in 2013. At the other end, Maharashtra and Andhra Pradesh have less than 10% of their government school students participating in the state-level test.

*Table 5.6* A comparison of participants from Bihar and Maharashtra

|  | 2012 | | 2013 | |
| --- | --- | --- | --- | --- |
|  | *Rural* | *Attends govt. school* | *Rural* | *Attends govt. school* |
| Bihar | 14.79% | 86.33% | 7.61% | 71.44% |
| Maharashtra | 52.53% | 9.94% | 56.72% | 5.84% |

If we couple this data with that of the rural participants, the complexity of the picture emerges if we compare the data of Bihar and Maharashtra from 2012 and 2013 (Table 5.6). The percentage of participants in the NTSE from Bihar are predominantly from government schools. Yet the rural participation rate is less than 15%. This suggests that students from government schools in urban Bihar are participating in the exam in good numbers. In contrast, an average of 54.6% of students in the two years are from rural Maharashtra. Yet the percentage of government school participants is less than 10%. This suggests that students from private schools in both urban and rural Maharashtra are attempting the state-level talent search in substantial numbers as opposed to those from government schools.

At the same time, we also need to understand the term 'government school' also encompasses a tremendous diversity ranging from the well-resourced Kendriya Vidyalayas (schools predominantly targeting the children of central government employees), Navodaya Vidyalayas (schools in every district for talented rural students), and special state schools like Delhi's Pratibha Vikas Vidyalayas to ordinary government schools and schools like the Kasturba Gandhi Balika Vidyalayas for girls from tribal areas.

Therefore, the participation rates of rural students and the data on the kinds of schools that they attend offer only a limited entry point into the kind of access to educational opportunities that these students possess. Nonetheless, one of the earliest and most routine registers of demographic analyses in the official documents of the examination is with respect to the predominance of urban winners in the examination. For example, consider this fig leaf from the 1969 Annual Report of the NCERT, which acknowledges and attempts to justify the overwhelming urban bias by referring to the 'nation'.

> Delhi students belong to several parts of India. It appears that either through Central Government Employment or business, many families from different parts of India has moved to Delhi. Consequently, Delhi is really all India and not a mere state of Delhi. The same reasoning applies to larger cities like Calcutta, Bombay, Madras, etc.

In this statement, one sees that the urban student is presented as somehow transcending her specific geographical context to embody the idea of an atemporal 'all- Indian' representative, irrespective of socio-economic status, caste, gender, or even regional identity (given the phenomenon of migration that is referenced here). We find a parallel construction of the absentee rural student. Similar to the construction of the talented student ('academically bright') as suffering at the 'altar' of the 'mediocres' and 'the slow learners' (discussed in the previous chapter), the 'talented rural children' were highlighted as a special target population within the general pool of 'talent' in need of special government interventions, separate from that which is available through the existing educational institutions. For example, consider this passage from the same 1969 Annual Report of the NCERT.

A preliminary evaluation indicated that the number of students getting the National Science Talent Search scholarships from rural areas is extremely small. On a close examination, it was found that whatever the technique adopted, the general background of a student goes a long way in his further achievements. Thus the atmosphere in which he lives and the school he attends do play some part in deciding his future. Since these additional scholarships are not forthcoming to rural students, they cannot come up. Under the circumstances, currently the Government of India is engaged in examining an additional supplementary scheme, i.e. scholarships for talented rural students. In this scheme, the idea is pick up talented rural children, give them scholarships and put in selected good schools. In addition, the scheme envisages a continued follow up of the progress of such students. It is sincerely felt that such a scheme will bring forth many rural children for successfully competing and winning the National Science Talent Search Scholarships.

(NCERT Annual Report, 1969)

Consider how the term 'general background' in the above passage is an umbrella phrase which obfuscates the socio-economic location and identity of the rural student, both of which are referred to through the phrase 'atmosphere he lives in'. The shortcomings in government provision are also similarly hinted at through the phrases 'the atmosphere in which he lives' and 'the school he attends'. However, these challenges faced by the rural student are completely individualized in a solution which smacks of a line from a recipe, i.e. 'pick up talented rural children, give them scholarships, and put in selected good schools'. One would not guess from this picture that the rural student (like the urban student) is enmeshed in a highly stratified society and the notion of educational advantage and disadvantage are intersectional and cumulative, drawing on the individual's caste, gender, and class position. The terms 'urban' and 'rural' as they are operationalized in the official literature of the NTSE do not tangibly convey the kind of access to educational opportunities that students possess.

## Participation and Success Rates of Girls

A survey of the number of male and female winners in the NTSE yields startling results, as observed below. There is a consistent under-representation of girls and the lack of any substantial improvement in the ratio (Table 5.7). To understand this statistic, we need to contextualize this data with respect to the representation and participation of girls at the secondary level (Classes IX to XIII) (Table 5.8). Firstly, only those students who have cleared Class X with more than 55% marks (originally at the scheme's inception, in science subjects) are considered eligible.

At the beginning of the 1960s, only 20.5% of the total eligible girls of that age group were studying in Classes IX to XII. By the next decade, there was a 5% improvement in their presence at this level of education. If we contextualize this information against the fact that the national numbers of participation in the first five years of the test (1964–1969) remained between 7000 and 9000, it is reasonable to assume that a very small percentage of girls actually participated in the test. The enrolment rates of girls at the secondary level have increased in the subsequent decades. But even so, the number of girls enrolled by 2010–2011 has only reached 44.7% of the total eligible population. At the same time, these figures have to be interpreted keeping in mind 'that enrolment does not necessarily imply attendance, which can result in an overestimation of females. 'For example,

*Table 5.7* Distribution of male and female NTSE winners

| Year | Total selected | Boys | Girls |
|------|---------------|------|-------|
| 1964 | 354 | 280 (79.1%) | 74 (20.9%) |
| 1984 | 750 | 653 (87.1%) | 97 (12.9%) |
| 2003 | 1000 | 773 (77.3%) | 227 (22.7%) |

Source: Report on the National Science Talent Search Examination (1964), Annual Report of the NCERT (1984, 2003)

*Table 5.8* Decade-wise percentage enrolment of girls

*Percentage of girls enrolled w.r.t total enrolment at various stages*

| Decade | I–V | VI–VIII | IX–XII/ Intermediate | Higher edn (Degree level and above) |
|--------|-----|---------|---------------------|------------------------------------|
| 1950–1951 | 28.1 | 16.1 | 13.3 | 10.0 |
| 1960–1961 | 32.6 | 23.9 | 20.5 | 16.0 |
| 1970–1971 | 37.4 | 29.3 | 25.0 | 20.0 |
| 1980–1981 | 38.6 | 32.9 | 29.6 | 26.7 |
| 1990–1991 | 41.5 | 36.7 | 32.9 | 33.3 |
| 2000–2001 | 43.7 | 40.9 | 38.6 | 36.9 |
| 2004–2005 | 46.7 | 44.4 | 41.5 | 38.9 |
| 2010–2011 | 47.9 | 47.1 | 44.7 | NA |

Source: Selected Education Statistics (2000–01, 2004–05, 2010–11)

if girls miss school more often than boys in order to do household reproductive work or cannot give as much as boys to their homework after school for the same reason, this means they have less access' (Saith and Harriss-White, 1999, p. 25).

Another aspect which also influences the performance of girls in competitive examinations like the talent search is that differences in the fields of education in which boys and girls are concentrated begin manifesting themselves from the secondary level onwards. Very often, in secondary schools in which girls are enrolled, science courses may not be provided at all – or if they are, they have less adequate infrastructure and less effective training than that provided to boys. It is noted that girls are actually directed to subjects like domestic science, handicrafts, and biology, while boys study vocational subjects or chemistry and mathematics (Saith and Harriss-White, 1999, p. 27).

In some cases, we notice state collusion in perpetuating this trend. In Uttar Pradesh for instance, with the adoption of the 10+2+3 system and a unified curricular approach, the state adopted a new scheme of studies in 1998. Under this scheme, all students are supposed to successfully compete in seven subjects. These include five subjects with no or little choice: science, social science, hindi, one more Indian language, and mathematics. While mathematics is compulsory for boys, girls have an additional option available in home science. What makes the situation worse is that majority of single-sex girls' schools in rural areas do not offer the choice of mathematics, the only available options are elementary mathematics and home science. In the case of two additional subjects where students have wider choices available, the single-sex girls' schools usually offer limited options of 'womanly' subjects such as sewing, cooking, etc. The scheme provides for a number of options including 'non-womanly' courses such as commerce, agriculture, accountancy, etc., but most of these are not offered in the majority of girls' schools. In this context, the chances of the success of girls in a test like NTSE are very small, especially when the NTS subject examination has a 20-mark section on mathematics, which cannot be answered by those girls who have opted for home science in the secondary level (Pandey, 2004).

Figures 5.5 and 5.6 examine the statewide differences in the performance of boys and girls in the years 1984 and 2003, respectively. We observe that in 1984, before the institution of the quota system between states, Delhi had the maximum number of female winners (49), followed by Tamil Nadu with 11 and Maharashtra with 9. The representation of girls among the winners is especially poor among the North East states, with only one girl winner from Assam. Himachal, Goa, Gujarat, and Kerala also didn't have any girl winners in 1984. In 2003, Maharashtra which produced the largest number of winners also produced the maximum number of female winners at 48, followed by Tamil Nadu at 21 and both Kerala and Karnataka at 18. The performance of the North East states taken as a whole in comparison to the rest of the country continues to be poor. Himachal, Jammu and Kashmir, Bihar, Tripura, Meghalaya, Mizoram, and Arunachal Pradesh had no female winners in 2003.

## Gender distribution of winners, 1984

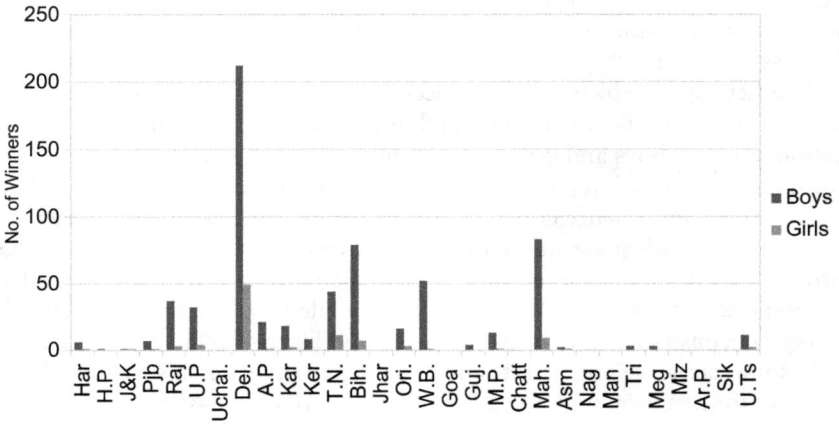

*Figure 5.5* State-wise distribution of male and female winners of NTSE 1984.

## Gender Distribution of Winners, 2004

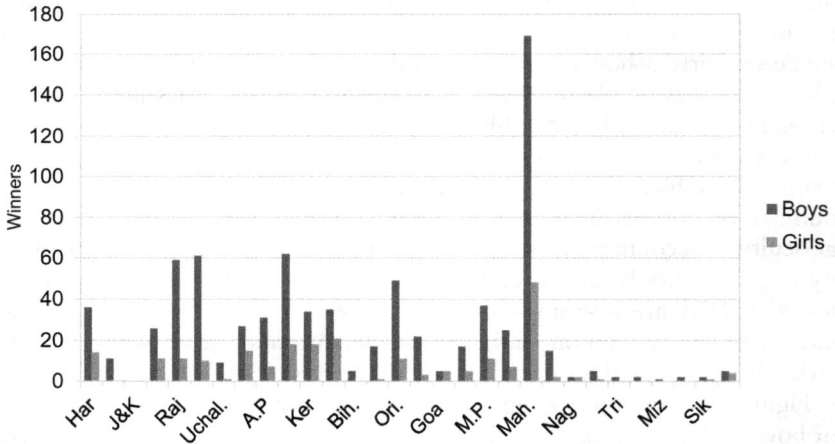

*Figure 5.6* State-wise distribution of male and female winners of NTSE 2003.

Therefore, one can argue that the low number of winners of girls who are produced by the NTSE is the result of the low rates of enrolment and completion of secondary education by girls. In addition, those girls who do appear for the test tend to be less prepared when compared to boys. This factors have to be interpreted with respect to compounding factors (in addition to the institutional ones discussed before), which limit the access of girls to secondary education in large parts of the country such as geography, caste, religion, ethnicity, age, and in some cases, disability, and prevent them realizing their potential.

*Participation and Success of Students from SC/ST Backgrounds*

The early reports of the NSTS do not carry a caste-based analysis of the winners and participants. However, internal studies of the talent search programme carried out by the NCERT in the late 1970s do acknowledge that winners tended to be from upper-caste backgrounds.

The NCERT Annual Reports of 1978 and 1979 mention that specific steps were taken to 'step up' the participation in these tests of 'rural children and children from weaker sections of society' to correct the imbalance in the award of the scholarship. These references seem to be a filtering of a greater attention to the educational constraints faced by students from Scheduled Castes and Tribes. For example, the phrase 'weaker sections of society' was policy-speak for categorizing members of Scheduled castes who tended to be rural, landless agricultural laborers and sharecroppers and was popularized by a national survey conducted by the Anthropological Survey of India between 1972 to 1975 on the 'weaker sections in populations in villages' of 'caste India' (i.e., excluding tribal societies) (Singh, 1993, p. 5). The mid-1970s also saw a nationwide study of the educational progress of SCs and STs, which while noting significant progress also mentioned that there was still a long way to go as far as their educational progress was concerned (Chitnis, 1981). Within the talent search programme, a tangible effort at inclusion was taken in 1981 when 50 additional spots were added to the national tally of 500 NTSE winners. In 1983, the total number of scholarships was increased to 750. Seventy-five scholarships began to be allotted to SC/ST candidates. Between 1985 and 1999, this number was again adjusted so that 680 candidates would be selected among all the Class X students appearing at the second national-level test and 70 would be SC/ST students. From 2000 onwards, the total number of scholarships was increased to 1000, with 175 being reserved for SC candidates and 75 for ST candidates.

Within the official literature on the NTSE (1964 to 2013), apart from noting the number of scholarships disbursed to the students from SC/ST categories, there is no further analysis of the role that caste plays in structuring educational opportunities. Unlike in the case of the analysis of the geographical location of winners and the state-wise quota in response or the attempts to articulate the advantages and disadvantages of urban and rural candidates or even the limited acknowledgment of the better performance of male candidates, internal caste-based analyses does not seem to have shaped the policy trajectory or design of the examination. Rather the reservation quotas seem to be implemented as part of a top-down policy process from the Ministry of Human Resource Development.

Within the data that is reported in the annual reports of the NCERT, one can see the following patterns in the state-wise break-up of the number of SC/ST winners. This is presented in Figure 5.7. We see that in common with the total number of winners, the maximum number of SC/ST winners is from Maharashtra (474), followed by West Bengal (180), Karnataka (169),

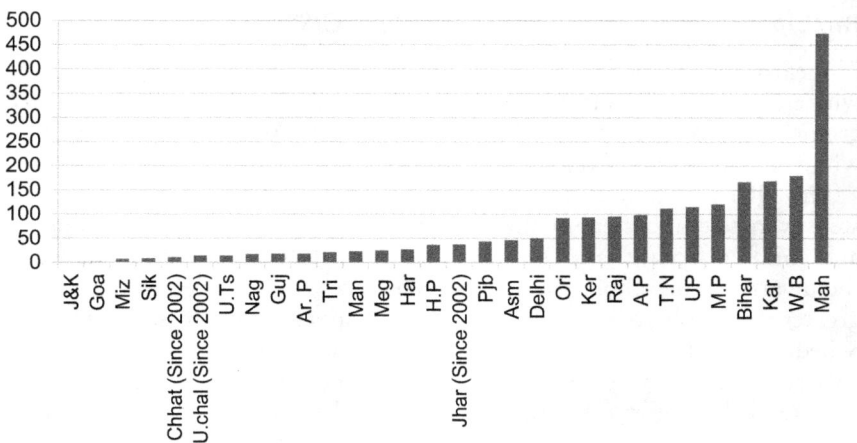

*Figure 5.7* State-wise distribution of SC/ST winners, 1985–2004.

and Bihar (167). What is notable is that the number of winners from Maharashtra is more than double the next set of leading states, which is similar to the pattern observed in the case of the total number of winners.

At the other end of the spectrum, Jammu and Kashmir have not produced even one SC/ST student, followed by Goa (1), Mizoram (7), and Sikkim (8). The case of Goa which has produced 60 winners between 1985 and 2004 and that of Jammu and Kashmir suggest that these numbers have to be contextualized in the light of the proportion of SC/ST students to the total school-going population of Class X in the state. This is important to note that to an extent these numbers represent the marginalization and underdevelopment of students from these communities within the state in general.

## Discussion of Findings

The nationalism which was present in the science talent phase of the programme primarily drew upon the belief that the capacity to produce science becomes an indicator of national progress, modernity, and even superiority on the international stage. However, the idea of the nation which underlies the second phase of the talent search programme is a less publicized version but one which reflects the mundane, procedural strategies of the state to continuously create and sustain national consciousness.

Geography underpins the national imagination in the conception and implementation of this phase of the NTSE. At the procedural level, the logic of 'coverage' and the logic of 'national' are conflated by using the federal model of the Indian union (the centre–state relationship) in the relationship between NCERT and the liaison organizations at the state level in the creation and administration of the examination, maintenance of records, and the conduct of nurturance programmes. For example, the decentralization of the examination from 1985, creating a preliminary state-level examination with fixed state quotas at the national exam, is a procedural attempt to ensure the representation of candidates from across the country at the national-level test and prevent the monopoly of places like Delhi, West Bengal, and Maharashtra, which were the top producers of winners prior to 1985. The procedures of the test build a certain idea of the nation through the values that they espouse.

The revisions in the scheme post-1980, which acknowledge the role of social contexts and backgrounds in shaping success in the exam, also signal 'inclusion' as an important value of the talent search. These include reservations for candidates from scheduled caste and scheduled tribe backgrounds since 1980 as well as for physically handicapped students (from 2008). From 2005 onwards, the purpose statement of the talent search as articulated in the NCERT Annual Reports repeatedly emphasizes the identification of talent from diversified groups like 'rural, urban, boys, girls, socially and economically disadvantaged, children with special needs and school dropouts'. Therefore, the idea of the nation, which is built through the procedures of the test ostensibly, seems to be tied to a democratic and equitable vision of the nation.

The data presented in this chapter interrogates this claim that the search for talent is a 'national' one and that the winners represent a 'national' cohort in their diversity. For example, the division of the NTSE into state-level and national-level examinations ostensibly seems to have created a more 'national' (in the sense of representative) distribution of winners. However, when the data is disaggregated at the state and district level, we see that there are huge skews. Post the establishment of state quotas for the number of winners sent at the national-level test, Maharashtra emerged as the major producer of winners between 1985 and 2004. It produced more than twice the number of winners coming from the next best-performing state, i.e. Bihar. Even within states, the distribution of winners does not always correspond to a uniform state-wise distribution. For example, only 14 states have at least 50% of their constituent districts producing at least one winner between 2001 and 2005. Twenty districts of the country (topped by Bengaluru Urban) have produced 40.2% of the total number of winners during this period. There is also a consistent under-representation of the number of girls among the winners. More than 75% of the winners each year continue to be boys. Similarly, most of the winners tend to be from upper-caste backgrounds. Despite the reservation of 175 seats for SC candidates and 75 for ST, those who are winners of the scholarships tend to be concentrated in a few states. Maharashtra produced nearly three times the number of SC and ST winners as compared to the next best-performing state, i.e. West Bengal.

From the test taker's perspective, 'national' signifies eligibility. A student within the specified age category anywhere in the Indian union is eligible to write the test. Indian citizens outside the country are also eligible to write the test (directly at the national level). However, we see that less than 5% of the total number of students who were enrolled in Class VIII wrote the state-level Talent Search Examination in 13 of 15 states in 2010 and in 14 of 16 states in 2011. In 2011, 2012, and 2013, less than 50% of the participants in the state-level test of six out of ten states were from rural backgrounds. Even in the case of several states with very encouraging numbers of rural participants, the problem is that these figures do not translate into the numbers of winners produced, who overwhelmingly tend to be from urban backgrounds. The kind of school (government or private), its location, and the kind of teaching–learning support infrastructure it possesses are the other significant determinants of whether a student has the information necessary as well as the confidence to participate in a test of this nature.

The absence of engagement of the test with the realities at the school level has prevented it from understanding the intersection of location (e.g. urban/rural), gender, caste, and class as creating structural constraints for a large number of students which impede their access to educational institutions and the ability to derive the full benefit from them. In the second phase of the examination, the emphasis on the psychometrics of the examination (for better identification of the ability of students) has continued to reinforce the construction of 'talent' in relation to success in the NTSE as inherent, culture neutral, and amenable to quantification. This means that the intent of equity in the talent search remains at a formal, rather than a substantial level.

Even so, the formal has a peculiar value of its own. These formal procedures are instrumental in establishing the ambit of the talent search as being conterminous with that of the nation's boundaries and the nation's democratic values. In this context, both those who design the examination and those who participate in it contribute to the sustaining of a 'national imaginary', i.e. 'patterned convocations of the social whole...[that] provide largely pre-reflexive parameters within which people imagine their social existence' (Steger and James, 2013, p. 23). Such procedures are part of the several symbolic forms of signification, repetition, and performance through which the state continuously affirms the unity of the people and their permanent identification with the nation as something that has already been accomplished (Bhabha, 1994). India's survival as a nation has been dependent upon this kind of quiet and incessant work performed through state institutions like the NCERT. This kind of symbolic work performed by India's state-nation (Stepan et al., 2010) affiliates citizens to the larger political community, while damping down differences from the larger 'national' narrative experienced by several communities within the country's territorial borders that possess other cultural and regional histories.

## Note

1 **Northern Region:** Haryana, Himachal Pradesh (UT from 1950–1971), Jammu and Kashmir, Punjab, Rajasthan, Uttar Pradesh, and Delhi. Though Delhi does not possess full statehood, it has been treated as such because of the significant numbers of winners who have originated from here.

**Southern Region**: Andhra Pradesh, Karnataka (erstwhile Mysore till 1973), Tamil Nadu, and Kerala

**Western Region**: Goa (post 1987), Gujarat, Madhya Pradesh, Chhattisgarh, and Maharashtra

**Eastern Region** (excluding the North East): Bihar, Jharkhand, Orissa, and West Bengal

**North Eastern Region**: Assam, Arunachal Pradesh, Manipur, Tripura, Nagaland, Meghalaya, and Sikkim

**\*Union Territories apart from Delhi**: The Union Territories, which later became states, were tracked from becoming as separate from the current set of UTs in the 1984 onwards. The remaining UTs have been taken as an aggregate, despite their regional status, because in some years, the break up of winners from each of these territories is not given. However, Chandigarh and Pondicherry are notable for the number of winners that they have generated.

# References

Bhabha, H. (1994). DissemiNation: Time, Narrative and the Margins of the modern nation. In *The Location of Culture* (pp. 199–243). Routledge.

Chitnis, S. (1981). *A Long Way to Go*. Allied Publishers.

NCERT. (1969). *Annual Report of the NCERT 1968-1969*.

Pandey, S. S. (2004). *A study into the causes of poor participation and achievement of girl candidates of Uttar Pradesh at National Talent Search Examination*.

Raina, M. K. (1984). *Background of the Talented. Delhi*. NCERT.

Raina, M. K. (1989). *An Intensive Study of the Accomplishments of National Science Talent Scholars*. NCERT.

Saith, R., & Harriss-White, B. (1999). The Gender Sensitivity of Well-being Indicators. *Development and Change*, *30*(3), 465–497. https://doi.org/10.1111/1467-7660.00126

Singh, K. S. (1993). Introduction. In Surajit Chandra Sinha (Ed.), *Anthropology of Weaker Sections*. Concept Publishing Company Ltd.

Steger, M. B., & James, P. (2013). Levels of Subjective Globalization: Ideologies, Imaginaries, Ontologies. *Perspectives on Global Development and Technology*, *12*(1–2), 17–40. https://doi.org/10.1163/15691497-12341240

Stepan, A., Linz, J., & Yadav, Y. (2010). The Rise of 'State-Nations'. *Journal of Democracy*, *21*(3), 50–68.

# 6    The 'Scholar' in the Search
## Memories of the Talent Search Programme

'*Well, it was more than thirty years ago… what can I say about it now?*'
Anil, a professor at IIT Delhi and a scholar of the 1980 batch, was the first
proverbial needle that I had unearthed from an online haystack of probable
National Talent Search Examination (NTSE) winners. His words were
daunting, to say the least. A variation on this theme also greeted me in my
next interview. Manas, yet another professor at IIT Delhi and a scholar from
the batch of 2000 (the youngest interviewee in my sample) elaborated,

> Quite honestly, it was fifteen years ago.. and since no one has ever asked
> me about it, I don't remember too much about it. The faint recollection
> that I have about it is that it was quite different from our regular tests.
> It was more like how you would consider an IQ test. But it was a little
> more deeper than that. It was…quite honestly, I have very faint recol-
> lections about it.

The specific connection that Manas made between being asked and being
able to remember is a point that memory researchers highlight. For example, in Conway's constructivist account of autobiographical memory,
'memories are transitory mental representations constructed by a centrally
mediated complex retrieval process. When a memory is to be constructed,
the retrieval process repeatedly accesses the knowledge base with respect to
current task demands' (Conway, 1997, p. 23). Conway's perspective of
memory as compilations of knowledge at different levels of specificity is
useful for us in providing a frame within which we can situate the narratives of the scholars. In this model, the knowledge stored in long-term
memory is understood as being part of three classes: the episodic, the
semantic, and the procedural. The first category encompasses memories of
experienced events, the second refers to types of factual knowledge such as
words and numbers, and the third to skills such as driving a bicycle and
typing. Of the three, the first is of interest to us because episodic memory is
considered interchangeable with what is called 'autobiographical memory'.
These episodes are vital to a person's understanding of one's self (Conway,
1997). Within the knowledge associated with autobiographical memory

DOI: 10.4324/9781003344902-6

are further layers, such as life periods, general events, and event-specific knowledge.

Of particular interest to us is the point that specific life periods may be associated with life tasks or problems which require special resources from the part of the individual. Such life tasks or problems tend to be very prominent during periods of transition. For example, in the shift from school to college, life tasks may be shaped by the broad themes of social and academic activities and may include aspects such as identity, intimacy, achievement, and power. Events from periods like this tend to be recalled in a highly emotionally charged register. Markus and Nurius (1986) observe that 'the themes appear to preoccupy first year college students are associated with the attainment of highly specific "possible selves" such as the self as a good student, the self as a competent individual living alone, the self as a socially attractive person etc'. These possible selves 'represent 'individuals' ideas of what they might become, what they would like to become, and what they are afraid of becoming and thus provide a conceptual link between cognition and motivation'. These are important because they also provide 'an evaluative and interpretive context for the current view of the self' (Markus & Nurius, 1986). For example, themes associated with the good student self might facilitate and prioritize the encoding of events carrying goal attainment knowledge for that self. After a retention interval (often of years or even decades), the individual may no longer recall the themes that characterized that period, although they retain access to the autobiographical knowledge structures created by the themes of the past 'good student' self. It is these long-term memory knowledge structures of the self that constitute a major part of the 'inventory of experience' (Conway, 1997, p. 28).

In my conversations with the scholars, a major reason for probing their experience of being examined through the NTSE and of being nurtured was to understand some meanings that they attributed to the label 'talent' and their interpretations of its impact on their lives. After my first two interviews, I adopted the practice of sharing a list of what I called 'talking points' with the interviewee at least a day prior to our meeting. Rather than the interviewee having to spontaneously recall experiences, I found that the interview discourse became far richer when they had some time to reflect on the topics. Several interviewees also wrote to me after our conversation because subsequent reflection triggered other thoughts. In the interviews, I generally used the first part of the meeting to discuss their memories of appearing for the examination. Questions like how they learnt of the examination; the questions they remember facing in the written examination, in the interview; regarding the project; and comments from teachers, family, friends, etc. were not only ways of triangulating the data that I was retrieving in parallel from official literature; my hope was that these would be an entry point in understanding how they fashioned themselves as learners. In the second part of the interview, I would direct the conversation to understand the perspectives of the scholars on the relationship between the ideas of 'talent' and 'nation-building'. (This content is taken up in the next chapter).

In this chapter, I draw your attention to salient points from the narratives of the scholars with respect to how they were selected through the talent search examination as well as their assessment of the programme's impact on their academic and professional trajectories. Two themes that frequently surface in the narratives of the scholars are addressed in greater depth at the end of the chapter i.e. the discourse of talent as a discourse of 'readiness' and the framework of one's 'self-concept' which is used in decision-making related to one's academic career/occupation.

## A Brief Profile of the Scholars

Eleven men and eight women, who were selected between 1964 and 2000, shared their experiences and reflections on the NTSE. Seven of them were part of this programme between 1964 and 1969, six scholars between 1970 to 1975, and five scholars between 1977 and 1984. Manas, the scholar mentioned above, was from the batch of 2000. His case is treated as an outlier, especially because he did not use the NTSE beyond a year. He had also cleared the Jagdis Bose National Science Talent Search (JBNTS), which he opted to avail instead. In situations where general insights are sought like socio-economic background, etc., his case has also been included. The names of the scholars in the following sections have been changed to protect their privacy. References to each scholar are usually followed by the year in which they won the NTSE in brackets.

Since details of how I gained access to these scholars as well as other sampling decisions were already discussed in Chapter 1, we can move to a quick portrait of the scholars. For the convenience of the reader, the separate profile of each scholar is attached in Table 6.1. The names of the scholars have been changed to protect their privacy.

### General Background

At the time of their respective interviews, 12 out of the 19 interviewees (63% of this small sample) were in academia. In addition, one scholar had started out in academia but moved to pursue an active career in politics and social activism. Six of these were women. This is significant because one of the purposes of the NTSE was to retain talented students in academics and research. Four were civil servants and three are associated with professions and functions related to the social sector (Table 6.1). At the time of the interviews, 14 were based in India, 4 in the United States, and 1 in the United Kingdom (Table 6.2).

Based on their location at the time when they were selected through the talent search, they represent a relatively diverse group hailing from the capital of the country and from eight other states (Maharashtra, West Bengal, Tamil Nadu, Karnataka, Gujarat, Madhya Pradesh, Kerala, and Rajasthan) (Table 6.3). However, as the previous chapter demonstrated, it should be of no surprise that 11 of the 19 winners were based in 4 metro cities: Delhi (6), Mumbai (1), Kolkata (2), and Chennai (1). The predominance of Delhi as a

*Table 6.1* Profile of scholars interviewed

| Winners (Name changed) | Year | Profession | Location during writing of NTSE | Gender |
|---|---|---|---|---|
| Shree | 1964 | Scientist | Delhi | F |
| Anjana | 1964 | Scientist | Delhi | F |
| Subha | 1968 | Scientist | Delhi | F |
| James | 1968 | Journalist/Professor | Bombay | M |
| Hari | 1969 | Scientist | Baroda | M |
| Arihan | 1969 | Scientist | Calcutta | M |
| Vinay | 1969 | Banker (Rtd.) | Pune | M |
| Krishnan | 1970 | Civil Servant (Rtd.) (IAS) | Ooty | M |
| Roshni | 1970 | Scientist (Rtd.) and Social Activist | Thiruvananthapuram | F |
| Rupa | 1970 | Social Scientist | Delhi | F |
| Suma | 1974 | Scientist | Baroda | F |
| Usha | 1975 | Scientist | Delhi | F |
| Malini | 1975 | Civil Servant (IAS) | Thiruvananthapuram | F |
| Dev | 1977 | Social Scientist & Politician | Shri Ganganagar | M |
| Mani | 1977 | Scientist | Chennai | M |
| Prathap | 1980 | Civil Servant (IPS) | Bangalore | M |
| Anil | 1980 | Scientist (Engineering) | Indore | M |
| Punit | 1984 | Professor, Education | Delhi | M |
| Manas | 2000 | Scientist (Engineering) | Kolkata | M |

*Table 6.2* Present place of work and residence of scholars

| Country | Number of scholars | Cities/regions | Male | Female | Total |
|---|---|---|---|---|---|
| India | 14 | Delhi | 2 | 2 | 4 |
| | | Mumbai | 1 | 1 | 2 |
| | | Chennai | 1 | 0 | 1 |
| | | Pune | 2 | 0 | 2 |
| | | Allahabad | 0 | 1 | 1 |
| | | Thiruvananthapuram | 1 | 3 | 4 |
| USA | 4 | Chicago | 0 | 1 | 1 |
| | | Laurel, Maryland | 1 | 0 | 1 |
| | | East Lansing, Michigan | 1 | 0 | 1 |
| | | San Francisco | 1 | 0 | 1 |
| UK | 1 | Cambridge | 1 | 0 | 1 |

city which has historically produced a large number of NTSE scholars is also reflected here. Eight winners hail from Bengaluru, Pune, Vadodara, Indore, and Thiruvananthapuram, cities that could be considered as the next tier below the metro cities in the 1970s and 1980s. Two subjects were from the towns of Udhagamandalam (formerly Ooty), Tamil Nadu, and Sri Ganganagar, Rajasthan. The urban–rural divide evident in the national level picture is also reflected in the urban origins of the 19 winners.

*Table 6.3* Location of scholars when they appeared for the talent search

| Origin | City | Number of scholars | Male | Female |
|---|---|---|---|---|
| Metro cities | Delhi | 10 | 1 | 5 |
| | Mumbai | | 1 | 0 |
| | Kolkata | | 2 | 0 |
| | Chennai | | 1 | 0 |
| Tier II cities | Bengaluru | 7 | 1 | 0 |
| | Pune | | 1 | 0 |
| | Vadodara | | 1 | 1 |
| | Indore | | 1 | 0 |
| | Thiruvanathapuram | | 0 | 2 |
| Tier III cities | Udhagamandalam | 2 | 1 | 0 |
| | Sri Ganganagar | | 1 | 0 |

Note: The classification of cities was based on the status of the respective city while the scholar was living there

*Table 6.4* Caste and religious profile

| Type of background | Caste | Male | Female | Total |
|---|---|---|---|---|
| Upper castes | Brahmin | 5 | 2 | 12 |
| | Kayastha | 0 | 1 | |
| | Khatri | 2 | 0 | |
| | Nair | 1 | 1 | |
| Other backward castes | Yadav | 1 | 0 | 1 |
| Christian | Roman Catholic | 1 | 0 | 3 |
| | Syrian Christian | 0 | 1 | |
| | Others | 1 | 0 | |
| Not disclosed | – | – | 3 | 3 |

The interviewees come from upper caste backgrounds, as Table 6.4 demonstrates. With respect to the employment of their parents, the majority of the interviewees had at least one parent who was employed in white-collar work (Table 6.5). The category with the predominant occupation of parents being academics and educators, followed by government service. Most scholars described their home background as middle class. Only four of them spoke of financial difficulties at home during their days of schooling and college.

The role of the family emerged as a significant topic in the descriptions of the scholars' pursuit of their ambitions and dreams. Eleven scholars have siblings with a minimum graduate degree. Thirteen scholars described their families as being equally supportive of both boys and girls in their studies. Eight scholars also described the key role played by their family members in creating and sustaining an interest in their subjects. Seven described the supportive role of their father as being a crucial component in their academic trajectory, while five attributed a greater role to their mother. Only three women scholars spoke of their families as being unsupportive in the pursuit of their dreams.

*Table 6.5* Profession of parents

| Employment of either parent | Description | Fathers | Mothers |
|---|---|---|---|
| Academics and educators | Scientist | 2 | 1 |
| | Economist | 0 | 1 |
| | Educator (college, school) | 3 | 2 |
| Government service | Banking | 1 | 1 |
| | Bureaucracy | 1 | NA |
| | Other government jobs | 2 | NA |
| Professionals | Journalist | 2 | NA |
| | Engineer | 1 | NA |
| Business | Owner | 1 | NA |
| Private sector | Employee | 1 | NA |
| Household management | Homemaker | NA | 4 |

*Table 6.6* Undergraduate academic choices

| Subject at undergraduate level | Total no. of scholars | No. of scholars between 1964 and 1976 | No. of scholars between 1977 and 2000 |
|---|---|---|---|
| Basic sciences | 13 | 13 | 0 |
| Social sciences | 2 | 0 | 2 |
| Engineering | 4 | 0 | 4 |

## Self Description as Learners

Sixteen of the 19 scholars described themselves as academically successful in school and acknowledged to be so by their teachers and peers. 'Topper', 'rank holder', 'good student', one who 'shone', 'excellent', etc. are a few of the terms which they used to describe themselves. Eleven scholars spoke of other awards which they received for success in their senior secondary examinations as well as other competitive examinations. Table 6.6 depicts the subjects which they chose at their undergraduate level. As previous chapters have noted, between 1964 and 1976, the scholarship could be availed only by students who choose to pursue an undergraduate degree in the basic sciences. Post-1977, it was opened to students who wanted to pursue a degree in medicine, engineering, social sciences, and management.

Eighteen scholars spoke about the choice of their discipline at the undergraduate level being driven by personal 'interest'. The one scholar who didn't was compelled by her father to abandon her dream of pursuing medicine and take up chemistry instead because of the financial incentive of the NTSE. This narrative of 'interest' as being the driving force for academic success is also bolstered by the fact that nine scholars (5 of whom belong to the period of 1964 to 1969 alone) spoke of the extra reading of journals,

*Table 6.7* Major influences on choice of academic discipline

| Influence | Number of scholars |
|---|---|
| Life and work of a scientist | 7 |
| School teacher | 2 |
| College professor | 2 |
| Parents | 2 |
| No response | 6 |

books, etc. right from their school days. This 'interest' is also presented as being nurtured at considerable cost and perseverance. For example, Hari, a scientist and a scholar of the 1969 batch, puts it across as follows:

> It was hard in those days; there was no internet and you couldn't just download. Now you can go to Youtube and listen to [lectures]. In those days we couldn't do any of those things. So you had to rely on going to the library and reading books. So I think I was influenced a lot by reading about scientists and science.

Table 6.7 captures the responses of ten scholars regarding the major influence behind their decision to pursue their choice of career. (The rest of the scholars did not speak of any special political or social figure or event which played an important role in shaping their trajectory.) As can be observed, seven scholars attributed the decision to take up an academic career because of the inspiration of the life and work of a scientist.

## Under the Examining Eye

What was it like to appear for the talent search examination? What details persisted in the minds of the scholars? For the purposes of identifying patterns in the experience and discourse of the scholars, the narratives have been analysed keeping in mind three time periods: 1964 to 1969 (the early years of the programme), 1970 to 1976 (the period which represents the mature phase of the National Science Talent Search (NSTS)), and 1977 to 1984 (the period from which NSTS became NTSE, open to non-science students as well).

### Learning about the talent search examination

Most of the winners (11) found out about the exam through their school (Table 6.8). Between 1964 to 1969, five (of 7) winners came to know of the exam through their school.

Shree (1964), a first-batch scholar, describes how the information about the test was specially given to some 'good students'. In other words, there was a tacit selection of students which was made by the school.

Table 6.8 Source of information about the NTSE

| Source of Information | Scholars between | | |
|---|---|---|---|
| | 1964–1970 | 1971–1976 | 1977–2000 |
| Information provided by educational institutions (School/College) | 5 | 5 | 1 |
| Peer group | 0 | 0 | 4 |
| Family | 1 | 3 | 1 |
| Former examinees | 0 | 2 | 0 |
| Advertisement | 1 | 1 | 0 |
| No recollection | 2 | 0 | 0 |

> ... We were told "Next week is the exam". We were selected; the so called 'good students' were selected and sent from the school. They said, "You guys go and write the exam. All the good students in science.... [In] those days, it was only science. So those of us, the top rankers or merit listers or those of who would the top ranks potentially ... we were all asked to go and write this exam, so we went and wrote this exam.

One's location made a big difference in whether one got to know about the examination. Hari (1969), who was from Baroda (Vadodara) in Gujarat, spoke of the lack of percolation of information about the examination in his place. It was his personal connections in addition to being from a family of academics, which made the difference in his case.

> [My mother] had a cousin who, I think, lived in Delhi and who is somehow involved in the government. He knew about the programme and he told my mother about it. And then she said, "You know, you want to consider doing this".... No one in my school actually even knew. There were a few people in Baroda who took the test. But I remember it wasn't very crowded... There were not more than thirty or forty people. It may have even been twenty or thirty.

Between 1970 and 1976, the school does remain the source of information for five winners. However, as the examination grew in prestige, other sources of information such as the family (for 3 scholars) and former examinees (for 2 scholars) also increasingly supplement the school. But even in the case of those who came to know of the examination through their families and their connections, one also sees the pattern of information about the examination percolating to those students who are perceived to be 'good'. Dev (1977) describes how information about the examination reached him in his small town in Rajasthan.

> It so happened that I did it very well in my class X. In that small town, I got a gold medal in State.. in those days, it was a great [thing] and in my small town, it was a big thing. That someone from that small town got a gold medal. So that's how people knew about me. That's how one

of them walked up [to me], the principal of the college where my father taught, [and]… said, 'Hello, why don't you apply for this one? You know, I saw this in the newspapers.' That's how, almost by accident, I got to apply… My school didn't know much about it, not even my teachers knew about it. [It] was very typical of a small town culture. So they said, 'This boy must give this'.

Two scholars came to know of the examination through its advertisement in various media (newspapers, posters in academic institutions, etc.). Below is the case of Vinay (1969), who came to know of the examination through an advertisement posted at his college in Bombay.

There was a poster put up on our college library, which said, 'Do you have a talent for science? Would you like to find out more about science?'. It was a very attractive poster… I was a science student and I decided to go for it. I applied for it.

### Preparing a Project

Scholars between 1964 and 1976 had to prepare a project as part of the selection process in the NSTSE. Their responses are tabulated in Table 6.9. We see two distinct themes in the way that scholars approached this task. The first theme that is dominant in many narratives is the idea of a 'self-directed endeavour'. The primary merit of the project is represented in the opportunity that it offered scholars to exercise their ingenuity, make decisions, and pursue ideas on their own initiative.

Execution of projects for the NSTSE

For the first batch in 1964, the project component was not compulsory. They were encouraged to take something which could be presented at the interview. It only became mandatory from the succeeding year till 1976. In Anjana (1964)'s account, her desire to do something based on her interests came up against the teacher's suggestion.

It was [in the] last moment [that I learnt that we needed to do a project]. I didn't have biology. I liked physics more, so I wanted to do a project in physics. My teacher suggested something, which I didn't like. In the end, I wrote something, but I did not make a project if I remember right.

*Table 6.9* Execution of projects for the science talent search (1964–1976)

| Scholars | Total | Male | Female |
|---|---|---|---|
| Those who submitted projects | 10 | 5 | 5 |
| Those who devised the projects independently | 5 | 3 | 2 |
| Those who admitted that they were helped by a teacher or mentor | 4 | 1 | 3 |
| Those who used research facilities outside the school | 6 | 2 | 4 |

Shree (1964), for example, spoke of doing a project on the human heart as part of her science club activities in her school in Delhi. Her interest in this topic was primarily spurred by her desire to be a doctor (till the NSTS forced her to change her trajectory). But she later decided not to take into the interview because she was not sure of defending it. Two other scholars spoke along similar lines, i.e. of the project as representing a continuation of things which they were pursuing on their own because of their interests. Arihan (1969) had read an article about making a telescope in the *Scientific American* and was already working on it at the Birla Industrial and Technological Museum, Calcutta, when he applied for the NTSE. As someone who routinely visited the British Council Library in Mumbai, Vinay (1969) also found his idea for the project in a book written by a contributor to the *Scientific American*.

What is common to all these scholars is that their interest in science could also be fostered by the facilities available in the metro cities of Delhi, Calcutta, and Mumbai. The importance of the scholar's location in terms of the kinds of resources that he or she had access to and in fostering a particular attitude towards science emerged in other narratives as well. Hari (1969), for example, explored this aspect at length.

> This was Baroda. Maybe Baroda was somewhat isolated. When I went to the summer school, a number of them were from Madras and they had done, they also did prescience … These guys went to good colleges, Loyola College, Christian College – Madras Christian College and so on. Their teachers were much more savvy and aware. They first of all told them about the programme and then they helped them with things. But some of them were second generation academics… [such scholars tended to be] a little more aware …. but there were also people whose parents did not have any connection with science. That was a good thing. But I think you have to be from the right area. Metropolitan areas.

Krishnan, a scholar (1970), who was from Trivandrum (Thiruvananthapuram) and who had to prepare the project under many constraints, represents his attempts in shades of a 'heroic' endeavour. (This is a frequent characteristic of representations of scientists in popular culture (King, 1999)).

> There was nobody to guide me for the project. So I had to make do with the materials available at home… [The ISC exams were over and I was waiting for the results]. I didn't have school at that time, or a college or a professor. I was in Trivandrum and I studied in Ootty. [There were] no telephones in those days. So no contacts. So I had to devise [the project on my own]. The project which I was working on was the electrical resistivity of certain electrolytes. …I was doing a lot of things on efficiency, basically. I had to do this thing. I had much bigger ideas but I couldn't get the kind of stuff….I had to go and buy

copper wires and plates, copper sulphate and all these kinds of electro-
lytes which you can not buy. Some of them are poisonous. They will not
sell it to you. Where to get all this and do this?... I had to do everything
at home. ... I had some twenty ideas at that time. ... [But] with the
materials available [I did what] I [could] do.

In contrast, many scholars admitted to taking help from their teachers or
from professors/mentors through inside connections. For example, Usha
(1975) did her project with a friend. Though their idea for the project was
their own, they also took the aid of a professor in IIT Delhi and approached
him for the use of lab facilities to complete their project. But in many narratives, one sees mixed feelings or some ambiguity in this status of having
been 'helped', which stems from cultural understandings of science as an
independent and original activity. Rupa (1970) was supported by her teachers but she emphasizes her passion, initiative, and independence. She validated these qualities by emphasizing her identity as a 'topper'.

> I still remember that. I was very excited. I went to the library, I went to
> the school library, I myself searched the books, where I could get some
> idea. Totally done by me. Of course I used to talk to my teachers also.
> They were the ones who informed me, but they wanted that I should
> work independently. [...]. I worked on rusting of iron as my project. In
> fact, I was the topper of my batch also in that year in 1970.

Roshni (1970) was forthright on the lack of originality in her work.

> My project was about Spectral lines... measuring... It was nothing very
> original.. it was ... I consulted a lecturer there... and she gave me a
> project... I just read up on it and did the experiments... it was...Now
> looking back, there was ... I mean, there was no originalness...

Suma (1974), goes into great detail of why she took help from a professor,
despite it being the 'traditional' thing to do.

> It was absolutely the first time that I had done any project. In school we
> barely had experiments. I was in a convent school...So maybe some
> professor mentioned it to us in class or something ...they said you have
> to do a project, and he suggested [my friend and I] should go to the
> chemistry department which was the one which traditionally gives the
> project. Now looking back on it, I realise it must have all been Hari's
> parents who must have started [the trend], so other people in the
> department have [also] come to know....

The scholar whom Suma refers to was also part of this study. Hari (1969)
was the first from this university to clear the NSTS, and his account sheds
light on how the giving of projects became institutionalized. His narrative

also reflects a mild embarrassment or self-deprecation about taking help for doing the project.

> [My mother] asked one of her colleagues in the biochemistry department if she would work on a project with me. She actually helped design the project. So I have to say, in hindsight, I got a lot of help with my project ... Because you know, I was in, sort of, the inside world. Of the two guys who followed me [and wrote the NSTS in the succeeding years], one was also the son of a faculty member. The other guy was my friend. He was from my school. He asked my parents if they could arrange a project for him and then you know, he ended up doing a project in that same department. So these were all sort of inside connections.

### A Different Kind of Examination

Chapter 4 'The "Science" in the Search' had touched upon the rationale among the test designers at National Council of Education Research and Training (NCERT) in creating a new kind of examination for the talent search based on multiple-choice questions. When the scholars were asked about their memory of the written examination (Table 6.10), six scholars described the test as a multiple-choice test, three as an aptitude test, and two as an IQ test.

Among the scholars selected between 1964 and 1970, three of them described the exam as 'novel'. Two found it 'interesting' as did another two who found it 'challenging'. The sense of novelty about the examination in the first decade also has to do with its multiple-choice format, which was

*Table 6.10* Descriptions of the NTS written examination

| Descriptions of the written exam | | Total | Scholars between | | |
|---|---|---|---|---|---|
| | | | 1964–1970 | 1971–1980 | 1984–2000 |
| In terms of type of test | 'Multiple Choice Questions' | 6 | 3 | 3 | 0 |
| | 'Aptitude test' | 3 | 0 | 2 | 1 |
| | 'IQ test' | 2 | 0 | 2 | 0 |
| In terms of curriculum | 'Outside regular curriculum' | 3 | 2 | 0 | 1 |
| | 'Thinking questions'/ 'Not testing memory' | 2 | 0 | 2 | 0 |
| In terms of personal experience | 'Challenging' (and synonyms) | 4 | 2 | 1 | 1 |
| | 'Novel'/'different' | 4 | 2 | 1 | 1 |
| | 'Not difficult' | 2 | 0 | 1 | 1 |
| None | No memories | 6 | 4 | 2 | 0 |

radically different from the essay type of questions that they were generally used to. For example, both Shree (1964) and Hari (1969) described their positive response to the test format. Shree put it as follows:

> 'I was very impressed with the exam itself, I still remember. As a young sixteen plus girl, this is fantastic, you are being tested for everything. It is easier to cover a broader area and it's not like 'woh saal mein aaya, toh is saal mein bhi'…At that time, I was very impressed with the multiple choice exam. We didn't have this multiple choice business in our time. All long essays and you know. In the long essays, the test cannot be done on a broad basis because you can only write so many long essays.

While admitting the quality of the questions, Krishan (1970) was more critical and drew attention to the kind of language used in the test.

> The questions were using American spelling……Everything was from the US. We could make out. It is not the kind of stuff you study in school. Some people said, 'They're making spelling mistakes; what is this, yaar[1]?' . But the quality was good. They were copying from something which had a high standard. But some could not understand the structure. The sentences were also .. problems which were alien to the Indian system of teaching. Even those who are good in the English medium and all that. [But this was] different. We have a particular way of formulating problems whereas theirs is different. It was totally copied.

Two scholars described the test as not testing 'memory' but rather 'thinking'. In Chapter 4, the distinction between the 'thought-type' questions of the NSTS as opposed to the routine memory testing 'factual' question was also a classification that the test designers had used. Roshni (1970) recalled these new questions in positive terms.

> I think there were three papers… three or two… one was factual questions… The other was thinking… the thinking part was the better one… I mean it was better for me… There was a lot that I did not know in the factual questions… But the thinking part was nice because it gave a comprehension passage and … you could figure it out… It needed some thinking… but it was nice… it did not need any information, you only had to think.

Dev (1977) connected this facet of the examination to its prestige. As he put it,

> …. it was not a test of memory, which is what differentiated it from other tests. And immediately gave it some respectability. So it was immediately more respectable to be an NTS holder than to have done

well in your class X exam or your class XII exam, because those you could mug up. Because it was not mugging, right from the beginning, that was one advantage which this test had, a certain prestige attached to it.

## Facing the Interview

Twelve scholars recalled their experience of giving the interview. The sheer detail in which many scholars were able to recall their answers to interview questions resonates with Conway's (1997) suggestion that individuals retain access to autobiographical knowledge structures created by themes of the past 'good student self'. For example, the memories which seemed to stand out for some scholars were the responses which they recalled as having impressed the interview board members. These responses were often presented as a turning point in the interaction. For Shree (1964), it was the answer to the question on who discovered the DNA helix structure. Roshni (1970) felt it was her answer of the value of Planck's constant that turned the board in her favour. Mani (1977) recalled the discussion that he had with two chemistry professors from IIT Madras on questions of atomic structure and energy production in the cell. Punit (1984) felt he impressed them with the Drake equation about the possibility of life on another planet. On the flip side, many still remembered the answers they got wrong. Anjana (1964) remembered *'goofing up'* the size of the atom and the size of the nucleus, while Rupa (1970) recalled at length the rationale behind her responses during the interview:

> I still remember one question they asked. How is a brick made? Eent kaise banti hai?[2] Now I know they are made from clay. Mitti se banti hain[3]. Clay bricks. We used to study cement in chemistry. [But] I thought yeh toh bada complicated hain[4]. I don't know what all they put into it. So I just said, 'I don't know'. I think that was the only question that I didn't answer. Otherwise I answered them all. I answered all their questions. It was a good grilling. Questions were there on the project as well as on general awareness.

Some scholars highlighted the role that 'personality' played in shaping the interview experience. Two winners spoke of their social skills that impressed the board. James (1968) expressed it as follows,

> I have very strong interpersonal skills and you know, I can dominate any damn group discussion, anything. So I remember in my interview, one of the interviewers saying, '[James], are you interviewing us or are we interviewing you?', because I had ended up interviewing the panel, you know.

Others were not so confident. For example, Anjana (1964), who had never attended an interview prior to this, described her experience as follows:

> The interview took place in the UGC. We were ushered into this huge room, with a huge table and around it were sitting all these studious looking people. And that was intimidating.... Without any clue, I just went in and saw all these people sitting around a huge table, throwing questions, trying to answer them... It must have been at least a dozen people there, I think so. A dozen people to grill one little school kid. Maybe they had psychs to sit and evaluate us'.

Suma (1974) spoke of being so shy that she completely blanked out during the interview.

> I was very scared and shy those days. Couldn't open my mouth, so I was hundred percent sure I am not going to get it...., most people expected of me to get through any written exam but the fact that I would flunk in an interview was also something expected of me. I think [the interviewers] must have been nice enough.

The aspect of language came up in two interviews. It was also another factor which contributed to Anjana's (1964) discomfort.

> You see, there was another thing, I at least was not very fluent in English at that time. So I was not very comfortable answering questions in English. That was there. I know that was something I carried into college. You know, so I wasn't entirely comfortable. This was a bit of a problem.

Dev (1977) echoed this. However, he insisted that he would only speak in Hindi and that was accepted by the board.

### The Role of Targeted Preparation

The issue of preparation for the examination was also explored with the scholars. Their responses are tabulated in Table 6.11.

Of the seven scholars who mentioned no preparation undertaken for the examination, five are from the first five years. The initial lack of preparation among scholars might have been tied to the novelty of the multiple-choice question (MCQ) format. As Shree (1964) put it,

> You see, when we went for the Science Talent Exam, we really were tested well because we had no clue about multiple choice. I mean we had never done multiple choice in our whole lives.

*Table 6.11* Preparation undertaken by scholars

| Preparation for NTSE | Total | Scholars between | | |
|---|---|---|---|---|
| | | *1964–1969* | *1970–1976* | *1977–1984, 2000* |
| No special preparation | 7 | 5 | 1 | 1 |
| School based coaching | 3 | 1 | 2 | 0 |
| Reading sample papers and commercial guides | 6 | 0 | 1 | 5 |
| Preparedness due to family's values and practices | 6 | 3 | 1 | 3 |

However, post-1969, there is an increase in scholars taking up some form of targeted preparation towards the examination. School-based coaching seems to have been the norm in the first decade of the examination. Subha (1968) and Usha (1975), both alumni of Lady Irwin School, spoke of the special sessions organized for the preparation of students for competitive examinations. In Subha's words:

> The NTSE was a very prestigious scholarship. And I was very lucky that I was in a school where they were giving a lot of emphasis on students who were good in science getting trained to appear for this exam. I was in Lady Irwin school… But when I was a student, it was like what is DPS today… And Lady Irwin was doing very well in terms of output in higher secondary exams. All competitive exams. Because they gave special emphasis … They would handpick some students and give them extra emphasis on extra-coaching and so on.

Apparently, these training sessions were also open to students of other schools (at least for a while). Rupa (1970), who studied in Queen Mary's school, spoke of attending Lady Irwin's training. Post-1976, there was an increase in the number of individuals using commercially available literature and sample papers in preparing for the test after 1969.

By the 1980s, it seems that the culture of MCQ questions as part of competitive examinations became more entrenched. For example, Punit (1984) spoke of practising multiple-choice questions for other engineering examinations and found that the test was fairly 'straight forward'.

> And for the national talent, I remember, there were a lot [of] kids doing it. But again, at that time, we were all looking for exams, 'kya exam de sakte hai?' You know. It was that way. There used to be these tutorial places and they used to have, like, camps by India Gate. Because we used to live really close to that, we used to live in Kaka Nagar…So we would just walk to those camps and they would have lots of quizzes and this and that and tests to practice, basically to sell their stuff … just practicing MCQ's and things like that.

Manas (2000) also spoke of preparing for the IIT-JEE, which was also help-
ful for an exam like the NTSE.

> Although there were books like <u>Prepare for NTSE</u> and books like that
> available in the market, I probably didn't… you know it happened at the
> same time as our pre-selection tests and all that. We were practicing for
> them anyway. We didn't do anything special… We may have looked at
> previous years questions and I looked at sample questions and all that.

Prior to the development of a coaching industry for the NTSE, the absence
of familiarity with the NTSE model of testing and school-based preparation, the family background of individuals seems to have made a difference.
Dev (1977), who wrote the exam a decade later, flagged the same issue,
when he described the exam as 'easy' despite having no institutional support or preparation in attempting the exam. He contextualized this response
by saying that problems which demanded logical and mathematical reasoning were easy for him because of his father who trained him to develop an
interest in mathematics and deductive reasoning.
None of the scholars attended or admitted to attending private coaching
classes. In fact, there seemed to be a great deal of emphasis in distancing
themselves from such an enterprise in two narratives. For example, when
I asked Anil (1980) if he remembered any preparation he undertook for the
exam, his prompt response was 'We didn't attend any coaching classes'.
Manas (2000) disclaimed even attending coaching classes for IIT-JEE.
(He referred to Bansal coaching classes in particular).

## The Impact of the Scholarship

This section documents the scholars' assessment of being a beneficiary of
the talent search programme. Apart from Manas (2000), all the other 18
scholars took up the award during their Bachelors' programme. Of these 18
scholars, 11 used the scholarship during their Masters' programme. Of
these, eight were winners before 1976 and three afterward. The reasons for
those who dropped out of the programme at the Masters' stage included
not qualifying the cut-off of 60% necessary to use the NSTE after their
Bachelors' degree (3 scholars, all of whom were from batches before 1976),
leaving the country (two scholars), changing from a science course to anoth-
er(one), taking a break in studies after the first degree (one), and getting a
job in between the Masters' programme (one).

None of these 11 scholars used the NTS scholarship in their Masters' but
used it to complete their PhD. However, all of them did do their PhDs, four
in India itself and seven abroad. Three scholars initially decided to do their
PhD in India but later they changed their minds. They went abroad. Two
were disqualified from using the scholarship because of a break which happened within their PhD. One scholar returned to do his PhD after working
for five years. He too could not use the NTS later.

## On their Academic Trajectory

Of the 12 scholars who were part of the Science Talent Search phase, eight mentioned that the programme's condition of pursuing the basic sciences made an impact on the choice of their undergraduate discipline. A recurrent theme in the narratives of scholars between 1964 and 1970 is the social expectation that a 'good student' would opt for science (pre-science or pre-medical) during their secondary level and then go on to pursue the prestigious discipline of medicine. In the totem pole of career options, basic sciences came after medicine. The specific impact which the NTSE had on the scholars between 1964 and 1970 is that it incentivized a career in the basic sciences as a desirable option in relation to medicine. What is common to many narratives is how the prestige and benefits associated with the NTSE matched their adolescent self-concept, which was an important aspect of their decision-making.

For example, consider Subha's (1968) reflections on the difficulty of making the decision to pursue an undergraduate degree in the basic sciences.

[My] medical admission was ensured for me because of my position in the higher secondary [merit] list. I would have gotten a scholarship of 100 rupees in a month. And then I got the NTS, the National Science Talent Exam. This was also a scholarship of 100 rupees a month. But it had more prestige because only a few [were chosen]. But even in the merit list, there were only twenty students. Somehow I was attracted to [NTSE]. It had a summer school attached to it and they covered your fees and gave money for books. But the only clause attached was that you had to take basic science. You can't take either medicine or engineering. So that was a little bit upsetting for me because I thought that I would automatically go for medicine.

Engineering, while being an attractive option, was less associated with a discourse of idealism and prestige, which apparently both medicine and the pure sciences possessed. Krishnan (1970) put it as follows.

I was selected in the IIT also. There was an issue because I was just 15 something not even 16 at that time. I wanted to be independent. The [NSTS] was giving 100 rupees a month. That was a lot of money in those days. [I] could be independent and could do upto Ph.D. ...In those days the value system was such that engineers were considered to be mercenaries-like. People who were very idealistic [and who] wanted to do something in the society, took pure sciences.

Krishnan's narrative also sheds light on the role played by adolescent impulses and desires in taking up the scholarship – the value of independence, idealism, and big dreams. However, one year into the BSc programme coupled with his experiences at the NSTS summer schools saw him reconsidering the impulse which made him reject engineering and take up basic science. He put it as follows:

> ...I was just 16 and half or something. It was my first year. It was a big shock for me. I couldn't come and tell my parents. 'You know, now I want to go back to engineering'. It was too late to try for IIT. You can't do last year's selection for next year. [My parents] would have said, 'You should have done this the first time! Now you want to interrupt this thing also!' They would have given me a lecture. To avoid that only, I went to do my second year.

Similarly, consider Suma's (1974) narrative. She also contextualizes her reason for opting for the basic sciences also in relationship to questions of identity (in particular, the identity of a topper), an issue of great importance during adolescence.

> For me, the [NSTS] made a real difference because in Baroda...I was considered a good student. People would say, "You've to go into medicine or this thing". I got into five year IIT [physics programme] and I was again very afraid, very shy, ...crying and all for a week before the [NSTS results]....I was feeling it was considered [that] it would be a 'come down' to do Physics in Baroda. It would not be considered very good. So when [the NTSE results] came, that made a difference because I was still toying with whether I should take engineering. In those days in MS university, I was the university topper. So you could get either medicine or engineering; you didn't have to give an extra test ...NSTS really helped, I felt that, ok, it's OK for me to do science because I have something which says...at this age you want to be differentiated from people who get into it because they could not get into engineering.

In Hari's account (1969), the language of destiny that he employs (i.e. 'a message') also points to the difficulties of decision-making in adolescence, including tensions between the consideration of one's sense of the self and personal assessment of capabilities versus the realities of admissions and so on.

> You know, [getting the NTSE] was almost like... I took it almost like a message. You know, I didn't get into the IITs and the there was only one medical school I might have gotten into, i.e. Christian Medical College, Vellore but it was mostly for women and they had only a small quota for men .. when I didn't get into those, I didn't know what I would do. Then I got the science talent scholarship, I thought it was a message and this is what I meant to do. I think it gave the slight incentive which pushed me in the right direction –basic sciences.

Secondly, among these narratives of the scholars, Shree's (1964) and Krishnan's (1970) accounts stand out for the negative impact which the choice of basic sciences made on their lives. Krishnan was more reconciled to it. However, Shree's narrative resonates with a powerful sense of 'thwarted destiny'.

> I mean, I was supposed to have been a medical doctor. (laughter). … Rama, Sheena, Manju, myself all of us, we were supposed to be doctors. Even Nandini, she is a doctor, she is in the United States, she is also one of the earliest ones who did not accept the scholarship, even though she got it…. But there are two other girls, Chandra, who is a doctor and Rama Gupta, who is a doctor. They didn't even write the exam. They said, 'We were told it was only for pure sciences'. But we said, 'Let's write the exam'. Then we decided to change our mind or trajectory or whatever…. I was really fond of biology, I was really good at biology, I got a distinction in biology in higher secondary. But my father said, if you want to do really good medical research, medically related research, you have to be really good in chemistry. Then I went in for Chemistry (Hons). My father died. Soon after, I got the science talent scholarship. I mean, I couldn't afford to go to college at that time. My parents didn't have the money to send me to college. I went to premedical and then when I finished premedical, then I switched when the science talent results came.…I think I would have done better research if I had gone into medicine because that was my primary interest. That was research which would have attracted me more. The unintended consequence of the Science Talent was that I didn't get [to do that].

In Krishnan's case (1970), the opportunity to attend the summer schools during his BSc as well as his experiences in the programme led him to a rejection of science as a career for him. In his Master's, he switched to another discipline, HR management.

### *The Experience of the Summer Schools*

A total of 13 scholars between 1964 and 1976 attended summer schools as part of the nurturance programme of the NTSE. Table 6.12 classifies five aspects which recur in the descriptions of the summer schools, including learning about new scientific developments, opportunities to handle new technology, mentorship by faculty, peer-interaction, and affective engagement. Each of these represents an important aspect of the exposure that students received through summer schools.

An unmistakable aspect of the narratives of the 13 scholars about their nurturance programme is the affective terms ('fun', 'the best', 'great') in which they recall this total experience. Such language can be found in the narratives of ten scholars (5 between 1964 and 1970 and 5 between 1971 and 1976). While a lot did depend on the institutions and individuals which organized the summer schools, there is some sense in which the scale or the nature of the programme itself was experienced as fundamentally different.

*Table 6.12* Significant aspects of the summer schools

| Significant aspects of summer school | 1964–1969 (7 scholars in total) | 1970–1976 (6 scholars in total) |
|---|---|---|
| Exposure to new scientific developments | 5 | 5 |
| Opportunity to handle new/ advanced technology | 3 | 2 |
| Understanding of the culture of research institutes | 0 | 1 |
| Greater awareness of options for Masters and Ph.D | 2 | 5 |
| Likeminded peer group | 5 | 4 |
| Inadequate peer group | 0 | 1 |
| First residential experience away from home | 2 | 4 |
| Description as 'fun' or pleasurable or in other positive terms | 5 | 5 |

For example, Subha (1968), herself the dean of a prestigious academic institution, surmised that:

> Actually the government had spent not just money but also infrastructure and effort in organizing summer schools. We [were] just a handful of students. Then a university had to take the responsibility of hosting them, taking care of their stay and then keeping them occupied. (Laughter). And make them feel that they learned something. So... when I look back...now I am in a position... if I am asked to do this, it is quite time consuming. (Laughter). But that time these people did it for us.

*Learning about New Developments in Science and Technology*

The summer schools introduced scholars to cutting-edge developments in various scientific fields through lectures, through field visits to research institutes, and through media such as science movies. In fact, the exposure to these areas of research during the summer school prepared the scholars in advance for the subjects which would be covered later in their universities. Arihan (1969) recalled the quality of the movies which were screened during summer school.

> There was a Chemistry, Chemistry Education Series, animated movies about different aspects of Chemistry. In the afternoons, there was often a feature on that as well which was included. Sometimes the movies were very very nice. [...] It kind of prepared you for the second year chemistry. So I was, we were already exposed to a lot of the things that we were going to learn in second year chemistry at the end of the first year.

A field visit of which she was a part endured as a fond and inspiring memory for Subha (1968).

> So one of the sight seeing trips was in Shimla. A new institute had started there, a Central Potato Research Institute… they were doing some agricultural research and trying to introduce new genetic approaches and so on… At that time, gene regulation was a thing which was coming up and people were studying it. It was a hard subject. A Nobel prize had been given in that field. We were not taught all that. But, so [this young faculty member at the Institute] started telling a little bit about it. Then we got so fascinated that we bombarded him with questions. We were walking from one point in Shimla to another point and all throughout the way, me and another friend of mine, the two of us would literally… you know like Pied Piper (laughter)… we were behind him and asking all questions and so on. So [those were] little opportunities that one got in the summer schools. I really valued that. It gave you, your mind, some food for thought. And those were the ones which helped, though they were small incidents, they really helped in understanding what can be done. Otherwise, botany, the way it was taught, I don't know what I would have done. I owe a lot to this scholarship for this reason.

Hari (1969) recalled the sum of these efforts as providing a 'feel' of things.

> They did try to expose us to some of the main ideas of 20th century physics. It was not necessarily the most modern physics, but at least, quantum mechanics, you know, what is the idea behind particle physics, what is the idea behind various things. So they tried to give us a feel.

*Learning to Handle New Technology*

A second aspect of the 'exposure' was the opportunity to handle new technology, i.e. literally 'getting a feel' for things. Vinay (1969) provided an interesting example.

> The person in charge of the summer school, one Prof. Gokhale, was very insistent that we should learn something of workshop technology also. So it included things like working on a lathe machine or working on a drilling machine. We had to do some projects also for that….there was a very good experience created. We had an x-ray diffraction machine, to do diffraction. And we went there even on a Sunday, a day we were supposed to have rest, myself and some three or four of my colleagues… So that was an entirely new learning experience… Because the sort of equipment we were exposed to, you know, it came much later in the university. Almost three to four years prior to what we were expected, we could handle equipment there. The X-ray diffraction machine, we never had any opportunity subsequently. It was only in the summer school that I could do it.

*Learning to Work with Scientists*

A third aspect of 'exposure' was the chance to learn from and interact with senior faculty at the elite research institutes where the summer schools were held. However, there were mixed reactions among the scholars about the faculty they were exposed to. Possibly, the quality of faculty associated with the summer schools had considerable variation. Apart from Subha's mention of the faculty at the Central Potato Research Institute, Arihan (1969) also spoke of the inspirational lectures that he attended.

> [In Chandigarh] .... I think we had some very very good teachers. It really enriched the chemistry that we learnt at that school.... There were the professors, there were four or five of them running the programme, they gave talks, they talked about their work which they were doing and we really had some advanced lectures, which really fired up our imaginations about what could be done. There was a talk on water which was really [fantastic].

In contrast, Krishnan (1970) mentioned his disappointment with the faculty at the summer schools in Madras and in Vishakapatanam. He tended to be cynical about their intentions and their capabilities. Commenting about his first summer school at Chennai, he noted as follows:

> [The summer school organizers were] all looking at NCERT as a source of funding. For them, NCERT would pay them pretty good for organizing this, for taking us out, for food and everything double or three times what regular students [were enjoying], getting paid. All the professors used to give me holidays for extra classes when they were being paid per hour so and so for looking after us. They were looking at us [like that]. Therefore, not the best guys were assigned to us but there was politics in the Department. So the head of Department decides who. People he likes, [those whom he] decides to help out for money and all that you know. He put those people there and not the really bright ones. Science is the area where you have to dissect all the time. You keep questioning. However big the person is, you have to keep questioning. The kind of sycophants whom they assigned to us, they were not really people who liked that kind of attitude....

Krishan is the only scholar among the 13 who held such a strong view about the politics of the programme. Vinay (1969) perhaps represents the middle, noting that the summer school programme did not really encourage intimacy between the faculty and the students.

> There was no special mentorship, you know, because we were a large group. Thirty to forty students. So there was no mentorship like that. And we couldn't maintain contact with the teachers for the simple reason that the technologies were not there. We didn't have internet or sms or things like that.

*Finding a Community of Like-minded Peers*

The fourth aspect of the exposure which the scholars received in the programme was the chance to interact with and live with peers from across the country for a month. Nine of the 13 scholars mentioned the presence of a 'like-minded peer group' was the greatest takeaway from this programme. The scholars mentioned how much they enjoyed the intellectual challenge which they received from one another. The residential aspect magnified this and created a very conducive environment for deep friendships. Subha (1968) put it as follows:

> It wasn't the teachers (in the summer school) who stimulated us at that time because again the teachers [of] Botany, Zoology had the same kind of notion of science [as taught in the university, with a great deal of emphasis of memorization]. But it was the other students and it was just that for one month, we are with a bunch of students who are very good and they can tell you about many things which you don't know, which you haven't read about. So I really benefitted from the Bombay contingent. There was this boy…He was a very different type. Most of us were very straight and we would do what we were asked to do. But he was a bit rebellious type. (Laughter). And in Bombay, the thing is that they were not just doing Botany, Zoology but they were doing Microbiology and then they were in touch with Tata institute of fundamental research. …But in that first year, he was exposed to quite a few scientists from abroad giving lectures in TIFR and he had gone there… He would tell us how he had learned this.. all very fascinating things and he would tell us why are you bothering about all this. This is not science. That was a good thing for us to feel the challenge. I was really missing that. I wanted to get an exposure, some direction into how one can do more interesting things in life.

Suma (1974) echoes a similar sentiment.

> Majority of the students [in school] were not interested in science as a career. So your peers were not motivating and it was really easy to be at the top of your class without doing any work at all. First time I went to summer school was in Pilani, I think there were eight girls and thirty-forty boys. It's the first time we [could] meet other people who are also interested in similar things. So that was very wonderful experience.

*Learning about Different Career Paths*

A fifth aspect was the 'exposure' to possibilities about one's future career trajectory. As Usha (1975) put it:

> One of the things that happened to people in summer camps was that… when they finish the BSc, by the time they get to some summer camps

in good places, they always knew that was the place they wanted to go back to do their M.Sc. That happened a lot. So there were many who joined IIT Madras or whatever. If they were from the south, they joined the IIT Madras programme. Then they went on to the US for doing their Ph.D.

Sometimes, professors at institutions directly invited promising candidates as well. Rupa (1970) provides an example of that (though she eventually did not take up this opportunity).

> All these summer camps gave me the opportunity to take up small research projects. I still remember. Although I was a person of Chemistry, in M.Sc previous, I went to CLRI, I worked in the Biophysics division. There was no distinction. The interdisciplinarity aspect was very much promoted. I worked in the biophysics department there on the electron microscope....I was fascinated by the prospect of working on the electron microscope which was in the biophysics department. So I said I want to work there. CLRI said, ok, go ahead. We have no problem. So I did that. Similarly in NCL Pune, I took up a biochemistry project, which was kind of a novel area at that time. They were so happy with my work that they said, why don't you continue here? This you can take up as your Ph.D work as well. I said I am not so sure about staying here. So they recommended my name to Indian Institute of Science, Bangalore.

### Financial Benefits

Among the accounts of the scholars, one sees the scholarship amount being considered as significant at least till 1980. However, the value of the money varied subjectively from scholar to scholar based on his or her class background and geographical location (Table 6.13). For example, James (1968) who did his undergraduate studies in Bombay, described the scholarship as 'an enormous amount of money'.

*Table 6.13* Spending patterns of scholarship money

| Type of expenditure | Total | Male | Female |
|---|---|---|---|
| Books | 8 | 3 | 5 |
| Savings | 4 | 1 | 3 |
| Tuition | 3 | 1 | 2 |
| Living expenses | 3 | 1 | 2 |
| Recreation | 2 | 2 | 0 |

At that time, it was an enormous amount of money. In 1970s, it was a huge amount of money. The first salary that I got when I started working in 1973 was 300 rupees. The scholarship, when I first got it, I think it was in 1969, was 150 rupees. In 1971, it became 500 rupees. Huge amount of money.

Arihan, who won the scholarship a year later and was studying in Delhi, described it as not a lot till he reached the PhD stage.

I mean, money always helps but it was not the determining factor for me … The money always came in one lump sum in the tenth to eleventh month of the year and there wasn't a whole lot of that lump sum to buy books or something like that… The book grant was pretty small. There was a hundred rupees per month which was an allowance and we used to get it at the end. That was [welcome] money. It could have covered more than the tuition, of course. Much more than the tuition. It wasn't certainly enough to cover the board. I was at Hansraj College, but I was at the University. … We had to pay the fees at the college and at the hostel, where we stayed at. We stayed at the hostel. [I used it] till the Masters at Delhi University. Then I did Ph.D. I started Ph.D at Delhi University and I was working on the Ph.D programme for two years, 73 to 75 and then 75 was when I felt (for USA)… At that time, the money was quite substantial. 350 or so per month and the contingency grant that would come for doing research. So that was good.

Similarly, Dev (1977), whose account we read about in Chapter 1, was studying in a small town in Rajasthan, described the amount as substantial even in the 1970s. A similar comparison can be made between two scholars of 1980. Anil (1980), who did his BTech at IIT Delhi, described his scholarship of Rs 200 as 'small' in the light of his expenses (His mess bill used to be Rs. 150.). On the other hand, Prathap (1980), who did his BA at Bangalore, described the scholarship amount as a 'prince's ransom', with 200 rupees being roughly equivalent to Rs. 2000.

Three scholars out of the total 19 scholars (two men and one woman) claimed that their scholarship money made a crucial difference in their pursuit of higher education. However, it must be noted that all three of these scholars came from middle-class families (based on the occupations of their parents), which valued education very highly. Therefore, it is reasonable to suppose that despite the economic hardship, all three of them would have found a way to pursue their higher studies even without the scholarship. As Mani (1977) put it,

I was one of 7 kids in my home and I would say that this was a major help for me in terms of IIT but I would have certainly joined IIT either way. But because I had the NCERT scholarship, it helped to make the simple decision and so I am always indebted to NCERT for giving me the scholarship.... I wouldn't say that [if I had not won the scholarship, I would not have pursued a career in engineering]. I would have still gone for engineering and still did all the things but it made my life much more easier.

Apart from the monetary value, two scholars (one man, one woman) spoke of a larger personal value of the scholarship. It added to their self-worth because they could be independent of their parents.

Table 6.13 represents the use to which the scholars put their money. Eight scholars used the money to buy their books (often over and above the book grant). Four scholars mentioned that they saved the money. Such savings were used to cover the costs of the PhD travels abroad. (Roshni's narrative (1970) stands out as against this trend. She saved her scholarship money and with her dad putting in an equivalent amount to her savings, she eventually bought a building. This building was changed into a community outreach centre for several poor people in her neighbourhood.). Apart from tuition and living expenses, two scholars also mentioned that they used the money for recreation (watching movies, etc.). There does not seem to be a discernible difference between men and women in their use of the scholarship money.

## Social and Emotional Benefits

One can classify the non-material benefits of the programme as encompassing social benefits and emotional ones. The social benefits associated with the award in the lives of the scholars were tied to its prestige and the kind of credibility it conferred upon them in front of others (Table 6.14). Fifteen scholars described the NTSE as a 'prestigious' award. But the value of the prestige manifested itself differently in the case of various scholars.

*Table 6.14* Socio-emotional benefits of winning the NTSE

| Socio-emotional impact of the award | Total number of scholars | Scholars between | | |
|---|---|---|---|---|
| | | *1964–1969* | *1970–1976* | *1977–2000* |
| 'Prestigious' award | 15 | 4 | 6 | 5 |
| Increased prestige in school | 4 | 1 | 1 | 2 |
| Increased prestige before peers | 6 | 2 | 0 | 4 |
| High value attributed by family | 3 | 1 | 1 | 1 |
| Increased personal self-confidence | 5 | 0 | 3 | 2 |

For those winners who are currently scientists or researchers, the award did not constitute a source of validation of their ability beyond the college level. The very nature of their field demands a constant demonstration of their worthiness to be part of the endeavour of research through publications, patents, and the projects or grants they received. Usha (1975) took up this same point with me after our interview in an email.

> I think what happens with people who gave up NSTS and moved into other fields is that this is the only award they have won in science. So their memories are vivid. For the others you should divide them into the super successful, the successful and the unsuccessful and then see how they think of NSTS. Of course success is a very subjective thing. When you have taken so many other exams, got other prestigious grants and awards, the NSTS' aura dims over time. But for all of us, it is what gave us our first stepping stone and sense of a community via the camps.

Post-1976, the contribution of the NTSE in the sense that Usha describes continued to diminish for those who pursued careers in science and technology. Punit (1984) expressed this as follows:

> [Being] a NTS scholar. I mean, it's still on my vitae. That's how you found me, right. So I am getting some credibility from that (emphasis) but I am not sure that, at least in my case as well as [for] lots of other cases, it truly did anything to… (long pause) enhance our careers in that sense.

Manas (2000) described the diminishing impact of the scholarship post-school. Even in school, winning the award did not entail special treatment.

> At that point of time, it was considered very prestigious but beyond that point, it sort of lost the impact….it was definitely a matter of prestige to be a scholar. But different kind of treatment probably was not there.

However, for scholars who did not take a research career in science, the award and its association with the pursuit of elite scientific research conferred a credibility which they drew upon even in their professional lives. Consider the following account of James (1968).

> When I stand up and when I give a talk somewhere, and somebody just says, '[Mr. James] is also a Science Talent Scholar', people take note of me. You can say 'I am an editor', but people wouldn't sit up and stand to that. I could say that I am a post graduate teacher, teacher of post graduate journalism but people wouldn't stand up to that. But when I say that I am a science talent scholar, then immediately, people sit up and take note of that.

Vinay (1969), a retired banker, also had a similar narrative.

> It made a difference' people you know, they would say he's a science talent scholar. People, you know, they were sort of, awed by me, presumably... Yes, [during the university time] and subsequently also. Even when I was working also. People were very much impressed. They would say, he has done M.Sc in Physics. Nuclear Physics was my M.Sc. And people would say he had got the Science Talent Scholarship.

Five scholars spoke of the personal impact of the scholarship in increasing their confidence and self-esteem. For example, Roshni (1970) described the receipt of the award in terms that were filled with affect. In part, her pleasure can be contextualized against her perception that her educational environments were mediocre. In that context, this recognition stood out.

> It's something I have been very proud of! Even now... I still write it in my CV, my bio-data. I remember, It gave me such pleasure to get it... such a wonderful news! I didn't know, I wasn't all that confident that I [would] get it.

By 1977, when the award was opened up for medicine, engineering, and the social sciences, the programme had accumulated substantial social capital from its association with elite science careers and this was transferred to the other disciplines. Dev (1977), who opted for the social sciences, described the impact of the award in terms of its contribution to his self-concept and confidence before his peers during his challenging postgraduate initiation into the culture of Jawaharlal Nehru University (JNU).

> NTS made a difference there. If I did not get NTS, something in the eyes of others as well as probably, in my own social confidence, may have been taken. My initial encounter in JNU was not a very easy encounter. There is a point of getting good grades and so on. And I started getting good grades. That's the only weapon I had, being from a small town. I didn't speak much English, ...I never needed to speak English in my life. English was a language that you read. You didn't speak this language. NTS...was the only way in which I could respond to this world. I could show them my A plus or grades or whatever. It was the only language that they could understand. Otherwise I didn't matter to them. In that difficult encounter, the NTS... the effect that NTS had was that it gave me self-confidence.

Another social science scholar, Prathap (1980) described the effect of the award in creating opportunities for leadership during his college days.

> ...they look to you for doing better in life as well as in academics... The principal and all would [nominate me] for some committees and editorial boards and things like that... such things which make a big difference.

Three scholars spoke of the award being highly valued by their families. As we noted in Chapter 1, Dev (1977) described the deep sense of accomplishment winning the award provided because of his father's approval.

## Discussion of Findings

This chapter provides an entry point to understanding the evolution of the talent search programme through the experiences of the scholars. Their experiences of being part of the selection process: preparing a project, attempting the written examination, and facing the interview along with the different aspects of the programme's impact on their personal and academic trajectories are the counterpoint to examining the same elements from the perspective of the test designers (described in Chapter 4). Before we move on to a discussion on the perspectives of the scholars with respect to the relationship between talent and nation-building in the next chapter, let us address two significant conclusions that we may draw from their narratives. Firstly, with respect to their experiences of writing the examination, there is a discourse of aptitude as 'readiness' that runs through the accounts. Secondly, for those who were part of the Science Talent Search Phase, the scholars' decision to take up the scholarship and pursue a career in the sciences reflects how occupational aspirations during adolescence are shaped by their self-concept. For these scholars whose self-concept had been reinforced as 'good students' in school, the label of 'national talent' either validated their preference for taking up the sciences in college or else helped them to overcome the social pressure to take up engineering or medicine.

### Aptitude as 'readiness'

An important theme which runs through the narratives of the scholars regarding how they learnt of the examination, how they prepared a project, appeared for the written examination, and faced the interview panel is that of 'readiness'. The scholars were either 'ready' or 'readied' to appear for the examination. Therefore, 'readiness' is closely connected to 'preparedness'.

When an examinee is successful (i.e. he or she correctly answers a question in a test), the conditions under which this knowledge was acquired are hidden. For example, it can be an educated guess, the result of familiarity with the area due to exposure and interest-driven reading or a trained response acquired through coached practice. Very often, these need not be exclusive categories. Yet the guiding assumption in an aptitude test is that it will uncover the innate skill and the knowledge acquired through interest and preparation. However, the conversations with the NTS scholars reveal that the conditions under which an individual is 'readied' to be tested produce two narratives of preparation.

The first type of 'readiness' is the product of an individual's submersion in a particular discipline or field of study. I use the word 'submersion' with its meaning of 'the act of descending below the surface' to indicate a deep involvement at the personal level with a subject or a topic. 'Submersive preparation' highlights individual agency and autonomy in acquiring knowledge. There is a greater prestige associated with such kind of self-directed acquisition of knowledge. For example, eight of the respondents spoke of extra reading (journals like the *Scientific American* or extra-curricular academic books) due to their interest in a topic. Of these, five attempted the NSTS before 1971, i.e. the first few years before commercial or school-based coaching became more prominent. Submersive preparation is presented as not being bound by the demand of being ready 'in time' for an examination. This kind of freedom from the constraints of time in the acquisition of knowledge is directly tied to the family background of the individual and his/her gender. For example, of the eight scholars who spoke extra reading, only two were women.

For a better perspective on 'submersive preparation', consider this narrative of Arihan (1969).

> I was four you know, when I decided I wanted to do research. (Laughter). It has nothing to do with that (He jokes in a laughing tone). I have a memory of that, I was such a mouthful. I used to play. I remember picking up some blocks and piling those scraps from an electrical shop just nearby where we lived. From their scrapyard I had picked up somethings and I was playing with them. I remember somebody asking me, (He mimicked a gruff scolding voice), 'What are you doing?' 'I am doing an experiment. I am making carbon dioxide'. Did I know what carbon dioxide was? I don't, I must have heard my brother or sister talk about it. It was such a nice mouthful of a word. (Laughter). It is an image from back then. So, yeah, no, I wanted to do basic research in science ....as far [back] as I can remember.

This example was offered by Arihan in response to my query as to why he chose to pursue a career in basic science. (This narrative is very self-conscious as to how 'his self' is presented. Its basis is possibly an interplay between his memory as well as representations by his family or others he internalized about his childhood self). In this narrative, the extremely young age of 'four' sets the stage for a representation of the individual's precociousness as well as motivation for the pursuit of science. 'Science' is even the object of his play and his modelling. His narrative demonstrates his awareness of science as being a self-driven and constructive activity. The reference to a family background where a child is freely allowed to play with scrap from an electrical shop and where there are siblings who familiarize him with a 'talk of science' points to the cultural resources in the individual's family. Arihan's father was a banker and his mother prized education very highly even though she did not possess a formal senior secondary degree. Class is definitely a factor, but so also is his geographical location. Their family location in Calcutta is significant because of its cultural valuation of scientists during that period and because of the proximity of their home to the Birla Science Museum, where this individual was mentored by one of the staff. His embeddedness within this socio-cultural matrix socializes him into a disposition which is receptive to the acquisition of knowledge, and it produces in him the readiness and competency for being tested later on.

The second type of preparation which produces 'readiness' to appear for examinations like the NTSE may be called 'specific preparation'. As the name suggests, this type of preparation is purposive and oriented towards a mastery of the logic behind a particular examination, so that the final responses of the individual resemble those produced by submersive preparation. James (1968) provided an excellent illustration of this approach, speaking of 'exam giving' as a type of habitus which is pronounced in some students. He compared his situation, with that of his friend Pavit's.

> There is a kind of student in the world who is a specialist in giving exams. So we are like exam givers, we are professionals. You can tell me to prepare for any exam …. It can be any damn subject, it makes no difference to us, as long as we understand what we are doing and we just top in that exam….We know how an exam is constructed and we know how to give that exam. How to excel in an exam. My contention to you …is that the NSTS is biased towards these kinds of students, who have been trained from childhood to give exams. [My friend Pavit and I] went for this examination. I was one of these exam givers and so I went through. Now Pavit is not an exam giver. He does not know how to give an exam. So he did not answer many of those objective type questions;…he did not know the 'technique of giving an exam', inverted commas. Not only that, he did not know the technique of giving the interview. Now I am also prepared. My mother was a teacher and my father was a trainer. I have been trained at home and trained at school to be a first rate interview person also.

James' account is notable because he attributes his mastery of specific preparation for examinations to unique conditions at home as well as at school. However, the same logic of preparation guides the coaching industry, which produces students who are 'exam givers'.

There is a greater prestige associated with 'submersive' preparation, which ostensibly signals personal interest as opposed to 'specific preparation' with a short-term goal in mind, such as that acquired through coaching or through association with an institution. It is especially hard for an examination to discriminate between the origin of the 'readiness' which is produced for examinations: whether it is 'submersive' or 'specific'. However, ironically, both types of preparation, which develop the dispositions for examination success either implicitly or purposively, are deeply tied to one's socio-economic background and gender.

Whatever the source of the preparedness or readiness, the significant aspect is the equation of this quality with talent or more precisely, aptitude. The relationship between these two ideas has been explored by the psychologist Richard Snow (2012), who traces the etymological root of the word 'aptitude' to 'aptness' in this sense of 'readiness'. He traces the connection of the word 'readiness' to other concepts such as 'concepts such as suitability (for a purpose or position), susceptibility (to treatment or persuasion), and proneness (such as accident proneness). 'The common thread through these and other related terms seems to be 'potential' or 'potentiality' – a latent, present, inferred quality or power that makes possible the development, given specified conditions of some further quality or power, positive or negative' (Snow, 2012). Snow traces the etymology of the word 'aptitude' to this notion of 'readiness' before it became gradually equated with intelligence and capacity in the sixteenth, seventeenth, and eighteenth centuries. It was then misrepresented and generalized as 'a single rank order of general intellectual fitness' for any situation' in the nineteenth century. It was subsequently appropriated in this manner by the mental testing movement in the twentieth century. Therefore, in these narratives of the NTS scholars, the discourse of 'talent' with regard to appearing in competitive examinations is articulated through the idea of 'aptitude', which captures the sense of 'readiness' and 'preparedness'.

## Self-Concept and Occupational Aspirations in Adolescence

The second main theme which one may identify in these narratives relates to the way in which the scholars assessed the benefits of the programme. A common theme in the discussion of their academic/professional trajectories and the enrichment through this programme (personal, interpersonal, academic, and monetary) is the significance of participating in it during their late adolescence. Therefore, the memories of the scholars demonstrate how various aspects of the programme are considered significant with respect to their engagements with 'self-concept' during that period. Linda Gottfredson's theory of circumscription and compromise in career choices (1996) provides several valuable insights on the 'self-concept', which helps us make sense of the points highlighted by the scholars. The self-concept refers to 'one's view of oneself, who one is' and has many elements, including 'appearance, abilities, personality, gender, values and place in society' (Gottfredson, 1996, pp. 183–184). Not all of these are equally central to one's sense of the self. While the self-concept is an object of cognition (i.e. 'me'), it also reflects the person as an actor (the 'I'). One of Gottfredson's key insights was that while people may not able to articulate their self-concepts and while self-perceptions need not always be accurate, people act on them and protect them. Especially in periods when one's self-concept is undergoing flux, as in the case of adolescence, this becomes a crucial aspect of one's decision-making.

An important aspect of socialization in childhood occurs in relation to 'images of occupations' (often called occupational stereotypes), and this includes 'the personalities of people in those occupations, the work they do, the lives they lead, the rewards and conditions of the work and the appropriateness of that work for different types of people' (Gottfredson, 1996, p. 184). These common images are organized into meaningful, shared cognitive map of occupations during childhood socialization. These maps are primitive early in life, but with increasing cognitive maturity, children come to perceive the same occupational map of the social order that adults do. As they move towards adulthood, adolescents 'identify the occupations they most prefer by assessing the compatibility of different occupations with their images of themselves' (Gottfredson, 1996, p. 184). Aspirations (self-concept) and vocational preferences develop closely in tandem, each influencing the other. Occupational preferences reflect an effort to both implement and enhance the self-concept. These are 'so tightly linked with self-concept because individuals are very concerned about their place in social life…Occupations are a major signal and constraint in the presentation of the self to society' (Gottfredson, 1996, p. 190).

A key point which Gottfredson makes is that occupational prestige is tied to perceptions about the intellectual complexity of the work. In other words, the prestige dimension is in effect a dimension of 'ability'. Additionally, one's most preferred occupations may not always be necessarily realistic or available. Many barriers may stand in the way of implementing them. Individuals therefore must also assess the accessibility of occupations when choosing which vocational alternatives actually to pursue. Occupational aspirations can be thus considered as 'the joint product of assessments of capability and accessibility' (Gottfredson, 1996, pp. 196–197). It is here that we can clearly see the contribution of the NTSE with regard to these scholars, especially in the first decade of its existence (from 1964 to 1976). The provisions and incentives of the NTSE made available a certain occupational conception associated with the basic sciences which was in line with the scholars' self-concept. In other words, the incentives of the programme mirrored what the scholars and society deemed them to be 'worth'. This was a major reason for the immense prestige associated with the programme in the first decade of its existence.

## Notes

1 A colloquialism, informally meaning 'friend' (e.g. as the word 'dude' might be used).
2 This is a literal translation of the previous sentence.
3 'It is made of bricks'.
4 'I thought it is very complicated'.

## References

Conway, M. A. (1997). The Inventory of Experience: Memory and Identity. In J. Pennebaker, D. Paez, & B. Rime (Eds.), *Collective Memory of Political Events: Social Psychological Perspectives* (pp. 21–47). Laurence Erlbaum Associates.

Gottfredson, L. S. (1996). Gottfredson's Theory of Circumscription and Compromise. In D. Brown, & L. Brooks (Eds.), *Career Choices and Development* (pp. 179–232). Jossey-Bass.

King, G. (1999). The scientist as pioneer hero: Hollywood's mythological reconciliations in Twister and Contact. *Science as Culture, 8*(3), 371–379.

Markus, H., & Nurius, P. (1986). Possible Selves. *American Psychologist, 41*(9), 954–969.

Snow, R. E. (2012). The Concept of Aptitude. In David Riley, & Richard Snow (Eds.), *Improving Inquiry in the Social Sciences: A Volume in Honor of Lee J Cronbach* (pp. 249–294). Routledge.

# 7 The 'Scholar' and the 'Nation'
## The Worldview of the 'Talented'

The modern term 'worldview' or the German word 'Weltanschauung' in its sense of 'a set of beliefs that underlie and shape all human thought and action' encompasses many directions in its intellectual history in philosophy and hermeneutics. It is variously understood as a 'symbolic system of representation', 'a frame of reference', or a 'coherent collection of concepts and theorems' that individuals use to integrate their knowledge of the world and the self into 'a global picture, one that illuminates reality as it is presented to [them] within a certain culture' (Aerts et al., 1994). There are two levels, connected but discrete through which worldviews may be understood. Firstly, worldviews have a collective aspect to them. They may be a collective set of ideas, which have been developed and theorized by 'an elite of philosophers, religious innovators, prophets, reformers, theologians etc' (Eidhamar, 2021) in religious and in secular traditions through which questions of ontology (model of reality as a whole), explanation (model of the past), prediction (model of the future), axiology (theory of values), praxeology (theory of actions), and epistemology (theory of knowledge). The second level at which worldviews can also be perceived and studied is that of the individual. The personal importance of worldviews lies in their power to shape the individual's ways of being, knowing, behaving, and sense-making. The construction of the individual worldview is a dynamic and ongoing process, shaped through socialization in daily life via practices, symbols, customs, and traditions. Notwithstanding its social origins, its biographical character also gives it a unique individual imprint (Eidhamar, 2021; Vidal, 2008).

The NTS scholars who were interviewed were those who were labelled as 'talented' by the state during their adolescence. This recognition was the precondition for a set of experiences and material privileges through the programme. A guiding assumption therefore has been that this life event can be a significant thread in their individual conceptions of 'talent', if one conceives of the concept as a tapestry woven through contexts and experiences. In my interviews with the scholars, I sought to use 'talent' as a frame of reference through which to understand a limited part of how the scholars made sense of their world, especially the relationship between state and

DOI: 10.4324/9781003344902-7

citizen. To clarify, this was not an examination of the holistic worldview of each of the individuals who have been selected.

Therefore, this chapter explores the responses to those interview questions which were designed around certain themes to elicit a deeper understanding of the scholars' views on 'talent'. The first half of this chapter explores the responses of the scholars to questions which use various aspects related to the National Talent Search Examination (NTSE) as a springboard to explore larger social concerns. The second part subsequently turns the attention to the design aspects of the NTSE and the scholars' critical assessments of the examination. In doing so, the larger aim of this exercise was to investigate the continuities and the disjunctions between the personal constructions of 'talent' and the 'nation' with respect to the institutional constructions at the policy and National Council of Education Research and Training (NCERT) level.

## Perspectives on Talent and Its Development

This section presents and analyses the perspectives of the scholars with respect to issues of talent development and its correlation with various social issues. During the interviews, I had asked the scholars to interpret some statistics of the NTSE in relation to larger social issues. Only one question was not directly connected to the test, i.e. the requisites for the fruition of talent. The remaining themes included the following: a) the non-identification of a large number of the talented through such tests, b) the persistent gender bias among the winners of the NTSE, c) the tendency of winners to be from urban areas, and d) the issue for reservation for SC/ST students in a talent search.

### *Requisites for Talent Development*

The scholars were asked to reflect and share the aspect which they felt had the maximum impact on the fruition of an individual's potential. I received 12 significant responses. Six scholars highlighted the interplay between the innate qualities of an individual and her environment. Four scholars gave the sole weight to the role of the environment in shaping ability. Only one scholar placed great premium on the 'inborn' traits of an individual, without which environmental influences do not count for much. Table 7.1 represents their responses. In the following section, the first two types of responses are discussed at length.

### *'Interplay between Environment and Innate Qualities'*

The six responses that highlighted the interdependence of the individual's innate qualities and environmental factors differed on how they weighted either the former or the latter.

Table 7.1 Scholars' requisites for the fruition of an individual's potential

| Most signficant aspect required for the fruition of an individual's potential | Total | Scholars between | | |
|---|---|---|---|---|
| | | 1964–1969 | 1970–1976 | 1977–2000 |
| Combination of innate qualities and environment | 6 | 3 | 1 | 2 |
| Socio-cultural environment | 4 | 1 | 2 | 1 |
| Genetic inheritance | 1 | 1 | 0 | 0 |

Arihan (1969) and Manas (2000) were two scholars who lightly touched upon the difficulty of pinning down the origins of talent through two distinct registers. Arihan (1969) treated his innate qualities as a 'blessing' reinforced by his passion and effort, the cultural context of his childhood, and his family environment.

> I was fortunate. I was blessed, I could say. Nothing of my doing. But because I had [aptitude], things came easy to me in science. I recognized that. But I was also passionate about it. I applied myself to it. I always pushed myself in that direction…. In general, at that time, all around, the idols were science. Scientists were the idols everyway or things like that. (Laughter). That was the culture that I grew up in…. Education was always valued very highly in our family.
>
> (Arihan, 1969)

In contrast, in Manas's (2000) explanation, no hint of the metaphysical exists. But the development of 'talent' is represented as scientifically explicable to a limited extent through predetermined genetics, the processes of neurological development, and the educational system in India.

> I am giving you a scientific answer to this. Probably [talent] is a mixture of genetic things and environment. Like everything else. A lot of academic talent or success, at least in the Indian system, depends on how well you remember. And in the longer term, if you look at higher studies, how well you can reason. The ability to reason and all that depends on how the wires in our brain are. How they are wired is both a function of genetics and how they are wired when we were learning things. So I think it is an open problem and I don't have the answer to it.

Roshni (1970), Krishnan (1970), and Punit (1984) acknowledged the importance of the environment and innate qualities. But they differed from Arihan and Manas, in that they gave precedence to one over the other. For example, Roshni (1970) spoke of how she was able to overcome the disadvantages she perceived in her education because of her personal qualities.

> Nurture versus nature... They both are needed... I think my environment was not that good. And in my career,... I have felt that...to some extent. But then ...I could always make up for it in some ways. As a scientist I had to interact with a lot of people abroad, I had lot of projects abroad... I had another fellowship in my career... the Alexander Humboldt fellowship. That exposed me to international level scientists... and nobody considered me stupid (chuckles)...In spite of having such a lack-luster education... not having gone to any premier education institute.

Krishnan (1970) used two metaphors to depict the relationship between one's personal qualities and the environment. The first is an agricultural metaphor that he deployed to articulate a point similar to the one made by Roshni. His argument is that talent will survive somehow or the other, despite the environment. The second is the image of the 'coolie' (a porter or unskilled labourer) to represent his perception of the general condition of stunted intellectual development in India. His implication that the opportunities for the 'talented' are menial and similar to manual work, rather than 'mental' work, also hierarchizes both types of work.

> It is like you plant a tree somewhere and the soil is very fertile and an ordinary seed will grow. But a special genetically developed seed will grow almost anywhere - even in the desert. [It] may not grow as much as it would with the required nurturing but it will keep growing. So environment is a very important thing....Freedom is a very important thing. Intellectual freedom should be there. Not cutting down ideas. Nonsensical ideas. We should have the freedom to think nonsensical ideas. From there you can make things realistic also. So that is not happening here. That is why we are a nation of coolies... we have money but all talents are going and doing coolie work.

Unlike the above two examples, Punit (1984) on the other hand, emphasizes that environmental advantages can overshadow the innate potential of individuals. In his case, he was conscious of the enormous difference that his class background as well as his location in Delhi made in securing access to resources that supplemented his passions.

> I [grew up] in Delhi, I was going to British Council Library, having access to all these things, which if you think of a kid from Meerut or a kid from where ever... (long pause) ... Jharkhand, would have no way (his emphasis) of competing with me. It's a completely unfair playing field. I come off sounding way more articulate and knowledgeable than

any innate smarts that I might have or passion or you know. I was very (his emphasis) passionate about learning. I don't, at all, want to play that down.

### 'Socio-Cultural Environment as the Key Aspect'

The socio-cultural environment of the child, without any reference to innate qualities of the individual, was highlighted by five scholars. Suma (1974) and Dev (1977) emphasized the role of the family in providing critical inputs which lead to the development of an individual's potential. For example, Dev's narrative (which I have reproduced at length) ruminates on the limitations of categorizing circumstances as 'advantages' or 'disadvantages'. He represents talent as the complex end product achieved due to a number of interconnected factors.

> In my family, I am a fourth generation teacher. My great grandfather was a Bachelor of Arts. So in 1890s, he had done a B.A and he had written his will in English language. This is how advantages travel. That is why even though I am a Yadav, I can speak English... At the same time, the fact that we are from an agrarian community tended to draw back. If my grandfather was a Syrian Christian ... or a Tam Bram, I would be sitting in Silicon Valley doing something there... The community pulled back the family's advantages... The disadvantages were that I lived in a small town...I was educated in a school, where, if I now look back, there wasn't much schooling in that school... Most of my friends were from agricultural families and went back to rural agriculture and so on... In my extended family, it was not a family where ... your academic things were held very high... And in terms of money, we couldn't afford very much. My father taught in a college. My mother taught in a school. .... But with all these disadvantages, I brought some very peculiar advantages of my nuclear family... [My mother] was so determined that her kids have to study... She was so single minded, like these famous Japanese mothers and education was the only thing that mattered to her. So that drive came from her. And my father... is one of those rare educated Indians ... who had allowed his education to touch his soul and transform him completely. Reading, writing, learning was his life. He read in order to transform himself... And this is something that none of my classmates had. So this one advantage that Mother was driven and Father was reflective, it outweighed everything else and gave me such an advantage over my peers that at the end of it, if I look more talented, it has to do with this thing.

In contrast, Rupa (1970) approached the question of what makes the most difference in the fruition of an individual's potential from an interventionist perspective. She laid great emphasis on the role of the school and its potential in facilitating the talent of individuals.

I think the school environment has a lot of impact. Most significant. For giving the exam, how were we informed about the exam? The school informed us. The school plays a crucial role in inspiring us. The school plays a crucial role in inspiring us. Today also, there are so many programmes running, run by Department of Science and Technology like INSPIRE and Children's Science Congress and all. Such information should percolate down from school.... The schools should facilitate the students to participate in the programmes. Sab se important link hain (This is the most important link)....You can't generalize home factors as such. There is so much variance. We have to depend much more on schools and teachers.

## The Non-Recognition of Talent

The scholars were asked to share the example of a person whom they considered talented but whose 'talent' was not recognized. Both the idea of who is a 'talented person' as well as 'the kind of recognition the individual deserved' were left open-ended for the scholar to interpret. The ten individuals who gave specific examples interpreted 'recognition' in different ways (Table 7.2).

### 'Inevitable Recognition in the Long Term'

Some shared an example with reference to not being identified by competitive examinations like the NTSE in school but who later went on to have successful careers. For example, James (1968) gave the example of his own wife, who despite not winning the scholarship, went on to win the Commonwealth Scholarship, do her PhD, and earn eight international patents. Suma (1974) mentioned the example of her husband, an exceptionally successful theoretical physicist and a winner of one of the most prestigious and lucrative academic prizes for scientific research, the Fundamental Physical Prize (Yuri Milner Prize). He was neither identified in the National Science Talent Search (NSTS) nor in the Jagdis Bose National Science Talent Search examination. Prathap (1980) gave the example of a fellow (unsuccessful) examinee in the NTSE, who later went on to become the principal of an international school in Bangalore. Punit (1984) described the inability

*Table 7.2* Types of examples given by scholars regarding non-recognized talent

| Types of examples | Total no. of scholars | Scholars between: | | |
|---|---|---|---|---|
| | | 1964–1969 | 1970–1976 | 1977–1985 |
| Specific example | 10 | 3 | 3 | 4 |
| Generic example | 8 | 2 | 3 | 3 |

of tests to capture the richness and diversity of interests and potential of individuals by giving the example of 'one of the smartest guys [he'd] ever met':

> In school, he was regarded as being smart but he was never, I don't think, given the due value by teachers [and the Board]....I don't think he was recognized for what he was. He ended up doing his undergraduate in Pilani, ending up going for his PhD in mathematics, almost completed that, then decided he didn't want maths as a career, went back to India, wrote a novel, an award winning novel with another friend on mathematics and truth, recently wrote a travelogue on Narmada. [He] became one of India's top respected journalists. You must have seen him on TV and stuff and very honest, very high level of integrity. That's why he gets fired from every job he gets. He won't compromise. He gives it left, right and centre. ... I think that's the thing, these tests are not looking at some of these things which actually count for so much, for success in the long term, things like integrity and character at all.

The key term is 'success in the long term', and in all the examples above, recognition of an individual's potential is treated as inevitable over time in various professional avenues, despite its non-recognition in the NTSE.

## 'Discrimination and Other Adverse Circumstances'

A second set of examples included individuals who could not attain the full measure of their potential because of a combination of social discrimination (gender and caste) and adverse personal circumstances (poverty, class background, and geographical location). For instance, apart from his wife's example, James (1968) described the story of his friend's sister.

> … the most intelligent out of all … five [of his siblings] was my friend and after him, was his younger sister. But because she was a girl, [their father] pulled her out of school at the age of 13–14… Especially in the rural areas, this is a dangerous time for girls. He pulled her out of school and he put her after the buffaloes. Take care of the buffaloes. That was her job… What did my friend's sister do? She got married, went to SriRampur, she joined the adult literacy courses there and she became one of the finest teachers of adults in the whole of Maharashtra ….She [had] a passion, she wanted to learn. Her brothers' books, she used to take and she used to finish all of that. Then she was after the buffaloes. Reading these books, she said 'just allow me to sit for the exams and I will pass'.

Dev (1977) provided two examples of the class and caste-based disadvantages that operate in rural India and prevent the fruition of the potential of individuals.

The first example that comes to mind is this classmate of mine... in school. His father was a darzi, a tailor. His mother was uneducated. He was brighter than I was. I didn't remember him very well in class VI, VII, VIII. But when I shifted in class IX, then I noticed him because he was number two after me. He did very well, smart boy and so on. As soon as he finished XI and he was the one who travelled with me for the NTSE interview in Bhopal. After finishing class XI, what did he do? He started working as a telecom operator. His father was a tailor and he... He thought that this was a smart, clever career move. To become a telecom operator after class XI. He got the job. He was pleased with it. I tried to argue with him, saying that this was silly. But bright as he was, he later [studied] and appeared for the Bank Manager's test. Now he is a bank manager. But if he had the kind of advantage that I had, and the kind of coaching, training... he most certainly would have been an IAS officer.

The second example that he provided was as follows:

This boy, he lives in my village, the house opposite mine. His father is from the barber community. He got some eighty percent plus marks in his class X. Think of a boy from a barber community, who goes to a government school, gets motivated on his own and gets eighty percent marks...he's an extraordinarily talented person. His ambition in life was to go to an engineering college. He has not yet gone to an engineering college. He finally went to an ordinary B.Sc. college. He struggled with his B.Sc. He has now come down to 65% in his B.Sc. and will most probably, be an unemployed boy; while had he some support, some encouragement, he might have gone for engineering

Hari (1969) also described the inability of the education system to respond to the needs of the individual with empathy, thereby leading to their exclusion.

There was one guy from pre-science whom I thought was reasonably smart. He came from one of the smaller towns of Gujarat. He was staying in a hostel. He had a lot of personal problems [and] anxiety and he may have been depressed... He was doing ok but he didn't show up for the exams. He had sudden panic [attacks]. When I went to the exam hall, I found this empty seat where he should have been.... That may be someone who would have benefited from counseling and I don't have any idea what happened.

In these narratives, recognition of these individuals' talent is a special kind of knowledge which is possessed by the scholars who narrated these stories. This is knowledge in hindsight, shaped by a reflection on what the scholars themselves perceived as advantages and disadvantages that helped in their lives. In the lives of the individuals described, the combination of socio-economic and cultural factors which led to the non-recognition of their talent has diminished their professional and occupational aspirations.

*'Unconventional Careers'*

The third type of example was of those who chose unconventional paths which do not have the kind of social status or privilege that careers like research or medicine or engineering possess. Suma (1974) described the trajectory of a fellow IITian:

> I had a classmate even in IIT who has chosen to go into women's issues… [When] people, who could have… perhaps if they had done the standard path of PhD, [they] would have done well in science… [choose] to do other things, … I would not call it a waste because I would think they were using their talents in different ways. I think those are also very important things to be done. For instance this person [now works with] women's issues… because she is a scientist, she brings a certain perspective to it which social science [may not be]…ready to look into those directions.

Here, the understanding of 'recognition' is that which exists among a community of peers in the scientific world. This cohort is represented as individuals who find it difficult to understand the career choices of those who reject this path. But the absence of rejection from such groups is not equated with a total lack of recognition. As is evident from Suma's narrative, such individuals are recognized in society by virtue of their 'renunciation' of personal advantage in order to pursue a larger social cause or good.

*'Wastage of Talent'*

Fourthly, there was an example of an individual who is implicitly accused of 'wasting' his talent because he would not court social acceptance or play by the existing rules. Roshni (1970) gave the example of an NTSE batchmate whose 'talent' tended to eccentricity.

> There was another guy from Kerala who also in the summer school… he was brilliant! One of the brightest people I knew… I mean nobody could come to him without recognizing him… but…but he messed up his own life to such an extent that… I mean he is not recognized in anyway… he's … he just… I mean he never had a career… he just sits in his house and… writes all kinds of funny things…which I don't think anybody reads… so, yeah, he was definitely the brightest of us…

In some ways, there is an underlying moral current in the presentation of the case of such individuals. Their lack of 'recognition' is in ways posited as 'justified' because of the lack of effort on the individual's part to direct his or her talent in socially accepted ways.

## 'Generic Examples'

There were also a number of scholars who provided generic instances of non-recognized talent. Such is the case of three classmates, described by Rupa (1970) as *'good, very good, very intelligent, bright'* (Rupa, 1970) but who were still unsuccessful in the NTSE. Anil (1980) gave the following example: *'IIT is full of talented people. But not all of them are NTSE scholars'*. These responses indicate a lack of engagement with what causes some people to be recognized in competitive exams as opposed to others.

Malini (1975), while offering a generic example, tied the issue of recognition to the kind of environment within which an individual finds himself or herself. However, she seems to suggest a certain 'fatality' or determinism against which individual effort seems futile.

> There are very brilliant people don't get acknowledged... If you have the innate potential to do [something and] you are put in the right environment, you will be able to achieve what you want to achieve. But if you are not in the right environment, if you keep on knocking the doors to find the right environment, by the time you are through the best part of your life, you've wasted it.

Vinay (1969) and Mani (1977) gave generic examples which suggested that individuals who are talented will be recognized if they put their efforts into it. Mani (1977) put it as follows:

> I think honestly, from school.... I think the people who got in, probably all were very good. In a city like Madras, people were well informed about it. If they were motivated and interested, they find about it. But in any field, it doesn't mean the NCERT or NTSE exams [will identify all the talented].

This again suggests a moral overtone similar to the examples regarding the wastage of 'talent' due to 'irresponsible' choices by individuals. The onus is on individuals, not institutions.

## Gender Bias in the Talent Search

Scholars were asked to comment on the statement that more than 70% of the winners of the NTSE each year tend to be boys. The responses fall into the following four categories: a) lack of opportunities, b) social pressure to be married and start a family, c) not a novel trend as this bias occurs in other exams as well, and d) the biology of difference between boys and girls. These are represented in Table 7.3.

*Table 7.3* Perspectives on the low success rates of girls in the NTSE

| Causes for poor success rates among girls | Total | Scholars between | | |
|---|---|---|---|---|
| | | *1964–1969* | *1970–1976* | *1977–2000* |
| Lack of opportunities eventually leading to poor success rates in such tests | 14 | 5 | 4 | 4 |
| Social pressure on girls to be married and start a family | 9 | 2 | 4 | 3 |
| The bias towards girls occurs in other competitive exams as well | 5 | 2 | 2 | 1 |
| Biological difference between boys and girls leading to difference in ability | 2 | 0 | 0 | 2 |

However, all these responses were contextualized by the scholars in relation to gender socialization, which creates different expectations about their potential and future. Two dominant themes emerged in the perspectives of gender socialization which the scholars raised. The first is regarding the beliefs about personal potential and possible futures in the minds of girls. The second pertains to how the socialization of girls leads to a denial of intellectual freedom and agency.

### 'Beliefs about Personal Potential and Possible Futures'

The reluctance of girls to take up science as a contributing factor in their poor performance in these tests was a recurrent theme in several narratives. Nine scholars analysed why girls do not tend to take up science or mathematics as a career and tied it to the social expectations, which shape their image of themselves and their future. For example, Manas (2000) analysed the socialization of girls as creating self-fulfilling prophecies or beliefs that these are subjects which they do not have the aptitude to undertake. He contextualized this kind of socializing culture as one of the signs of India's continuing 'backwardness'.

> For whatever reason, people think that girls are not good at science. It is completely flawed and in many ways our society is completely primitive. Especially when it comes to looking at girls. This is something I find completely ridiculous, but that is the truth. Also fewer girls actually take up science in eleventh and twelfth. That is why the denominator becomes much smaller and then from among them, as a girl, you are

told that you are not talented, you will not... That is still done. As of 2013 (the year of the interview), people still do that. This is not 1950. Why would she go for this talent exam? Even in the few numbers who are allowed to choose science, I would say that is a probably better phrase, even among them, whoever wanted to sit for this exam, had they appeared for the exam, you would have larger numbers. If you are constantly told, what is the use, then....

(Manas, 2000)

Another point that was raised was how the attitudes for success in competitive examinations are purposively developed in boys in the training they receive at home and in a peer culture which reinforces them. This argument, made by several scholars, is that success in these exams depends upon the fostering of a competitive spirit and that is tied to perceptions about one's possible destiny. Even the two scholars who spoke of the biological differences between men and women influencing their cognitive potential included a psychological and sociological dimension to their analysis. One of them, Prathap (1980), put it as follows:

Most of these tests test your cognitive skills. Research has proven that to some extent that men in terms of cognitive thing have higher potential. Brain potential depends on quantitative aptitude but women must have higher emotional intelligence aptitude potential but that is not tested in an exam like this. So even they talk in terms of brain size being different and the areas of the brain which are developed for the woman and the man being different,.. the conditioning of the women [is] not to be that competitive in Indian culture. [That] could be another sociological [factor]. There will be genetic reasons, there will be neurological reasons, brain study reasons and then it could be sociological reasons. So women are breaking out of the conditioning only in the last 30–40 years that they can do anything that the man has.

This is also a point brought up by Punit (1984).

The competitive thing comes very easily to boys. 'I got this in my exam', 'How much did you get?'....Social issues in India with girls too [play a part]... in terms of gender, assumptions about what you should study, what kind of a role in the future you should have, and yeah, you might study medicine but you are still going to be a house wife. That's going to be a primary thing. So there are all those gendered things going on at the same time....

In his narrative, James (1968) contextualized the NTSE trends with respect to the attrition of women in higher education, especially in rural India. He linked it to the socialization of girls for an early marriage, rather than paid work.

So there is in society a very deliberate discrimination. ... It is not just in NSTS.... You see the percentage of girls sitting for the 10th standard examinations ... the 12th standard examinations... the IIT entrance examinations. They are falling. Where are the girls going? The answer is very simple, Rachel. Because you are a woman, please appreciate this and understand this. Getting married off. They are being prepared for a life of marriage. They are being brought by their parents and being moved away from the world of work. Professions, wage earning work. Becoming unpaid wives and unpaid mothers and unpaid sisters and God knows what. So a small reflection of this, you will see in NSTS. Nothing unusal.

## 'Denial of Intellectual Freedom and Agency'

Among those who analysed the socialization of girls, three scholars described girls as being more compliant, docile, and un-original due to the denial of intellectual freedom during their upbringing. While this may not affect success in conventional examinations, it plays a part in boys performing better in tests which demand the exercise of analytical faculties and a questioning attitude. This also ultimately makes a difference in the kinds of research which men and women produce. Krishnan (1970) described the 'embodiment' of such attitudes in girls as follows:

> Society dominates girls much more than boys. Everybody is dominated. Society dominates girls much more. "Don't do this, do this, do that!". See, intellectual activity and scientific research is based on freedom. It is a rebellious streak in you if you say that, "[I] don't believe that is right. I believe it is wrong". That kind of extreme courage is a rare thing. Not everybody has it. One in a million has it. It is not nurtured anymore. It has suppressed a lot in girls. If girls and boys are treated equally I am sure that girls would... In engineering if a little bit of thinking is involved, boys do better. For medicine where there is a lot of repetition of knowledge thing, girls do better. Because they will sit down and study. It is all consequence of our social prejudice. Not because of any intellectual difference.

Roshni (1970) also linked the same to the lack of 'originality' in the work produced by girls.

> Somehow, by the way they are brought up, girls end [up] being less original than boys. Boys grow up with a lot of freedom to explore and rebel and... think on their own... I don't think they are intrinsically more intelligent. I definitely don't think men are more intelligent. They do, they are brought up in a different way... Freedom is very important and ...the stereotyping of girls mugging and boys thinking, is ... is kind of... true.
>
> (tentative, apologetic tone)

In the examples which were cited above, girls are represented as being passive in the face of patriarchal socialization (though the word 'patriarchy' is not used even by one scholar).

Two narratives, both by women scholars, lay the onus on girls for their performance in the NTSE. For example, consider this narrative of Suma (1974). While she acknowledges the differing socialization of girls and boys, she lays the responsibility on girls to take such examinations 'seriously'.

> Many of the girls are not serious. I mean they applied [for the NTSE] without being very serious...There is a lot of pressure on the boys, both from their peers and from their parents to succeed the same kind. Because these are very competitive, they have to study day and night and other things ... Girls don't perform [because] neither the parents nor the peers put enough pressure on them to say that "This is the most important thing that you have to do". I think the natural thing is to be lazy for everyone. There needs to be something to motivate them to go beyond that. So I think that is perhaps a little because of that fact and then, of course, everybody knows beyond a certain age, there is a lot marriage pressure. But I think [at] 15,16 17, more than pressure, girls are taking it easier because there is no one pushing them. It's only in those families where they treat the girls same as boys, that they push them [saying], 'You also have to study, you also have to excel, you also have to be the top'.

Suma does not take this line of thought very far because she essentializes the members of both sexes as being inherently 'lazy' and dependent on external stimuli for achieving goals. The difference in Subha's (1968) analysis, on the other hand, is that she critiques girls for not fighting or resisting the socialization that they receive.

> For one, I feel that girls are doing well in the board exams because girls are generally more methodical and also if a rigorous kind of thing is told, 'You go through it, you do it', they are able to cope with it better than boys. Nowadays they go through such aggressive coaching and that coaching helps them to get the marks. But the Science Talent Search is a little bit more than that. You have to develop even more analytic skills. For that, also, coaching helps. But some inborn thing also should be there. The way that girls are brought up in our country, I don't think many of them feel that they have to sharpen those skills also. They are quite happy to get the marks and clear the exam. After that they just push it away. I found so many girls- they take science, they do well and then they say, 'Please don't ask us to look at science again. Don't even tell their names'.

### Urban Bias in the Talent Search

The scholars were asked for their perspectives on why urban students tended to be more successful in the NTSE. Table 7.4 sums up their responses.

Table 7.4 Scholars' analysis of success of urban students in the NTSE

| Perspectives on the success of urban students in the NTSE | Total | Scholars between | | |
|---|---|---|---|---|
| | | 1964–1969 | 1970–1976 | 1977–2000 |
| Greater awareness and exposure | 9 | 2 | 5 | 2 |
| General trend in competitive exams | 3 | 0 | 1 | 2 |

*'Location and the Nature of Opportunities'*

A majority of the scholars (11) discounted that the difference in performance was a difference in ability between students of rural and urban areas. Nine scholars held that students from urban areas have more awareness and exposure. As Hari (1969) put it:

> Certainly each region has a lot of talent. I think the urban thing is more about opportunities and wealth and things like that.

Mani (1977) suggested that their expression of the ability and potential of rural students may manifest in ways which are different from that of urban students. A discerning test design would acknowledge this.

> For example, a person from the rural area may not know the latest in mathematics or physics but they may have a unique perspective on health issues or technical issues in their village ....I have often [seen] in websites very cool innovations [that] are happening in the rural areas of India. We never know, they may be some of the brightest people.

*'Institutional Correctives'*

Several scholars argued that one needed to examine possible solutions for this urban bias from an institutional perspective. Both Dev (1977) and Suma (1974) emphasized that the intersection of class and location disadvantages created an immense barrier for these students to be successful. Suma argued that only directed intervention can secure a future for those students with talent and aptitude in various areas.

> I would think... if you belong to a certain class of society [and are talented]... I would think that right now in the Indian middle class, if some students are talented, they would manage to make it. But I would assume that if you are from a rural area from a lower class...You see, [in the case] of HRI (Harish-Chandra Research Institute, Allahabad), we are having an elite institute in the middle of a village. It's like the

difference is so marked that it's kind of hard to see a talented kid from there [will make it in here]. How will they make it? HRI students ran some schools or tuition classes for these kids. So maybe if one of them gets recognized, then it may happen. Then it's still because HRI recognises it. Without recognition, those kids in villages and rural areas and all [would not make it].

Rupa (1970) emphasized the need to intervene at the level of the teacher in order to bring more awareness about the programme in rural areas. This lack of awareness was presented as a major factor in preventing students from these areas from participating in the exam.

This is a matter of concern to address. No matter how, children should get into [the talent search programme]. A lot depends on teachers. Not that we don't have good teachers in rural areas. We may have good teachers in the rural areas also. But the problem is that the teachers themselves may not have heard about the exam. So awareness of school teachers is a good strategy....If you want the students to be aware of the exam, first make the teachers aware. First sensitize the teachers. First train the teachers. Only if they are interested in projects and activities and all, then only they would be interested in helping students. But teachers have a crucial role to play. Teachers are teachers whether in rural areas or in urban areas. We have to make special provision so that rural areas get that support.

Hari (1969) suggested several levels to intervention to help rural students.

For people of rural India, [accessing opportunities] is very difficult. First of all, there are no libraries and good teachers don't want to go and teach in rural areas. It is a huge problem I think it will improve only when you have better inter-connectivity and not just internet. You have to commute and to travel easily. Transport needs to be improved and they should have proper accommodation so that people don't feel that going to teach in a small town or some rural area is like some sort of a prison sentence, a failure. Of course people won't go to a rural area if they don't belong to that region.

### Quotas in the Talent Search

Scholars were asked to comment on whether the cumulative historical disadvantages which students from SC/ST backgrounds experience should be considered in a talent search (Table 7.5). The NTS reserves 15% and 7.5% scholarships for SC and ST students, respectively.

The majority of the scholars (10) held that reservations are not the solution to addressing the disadvantages of historically marginalized groups

*Table 7.5* Scholars' perspectives on reserving scholarships in the NTSE

| Perspective on reservations | Total | Scholars between | | |
|---|---|---|---|---|
| | | *1964–1969* | *1970–1976* | *1977–2000* |
| Reservations are not the solution | 10 | 2 | 4 | 4 |
| Reservations are necessary | 5 | 2 | 1 | 2 |

like the SC/ST students. Only five scholars endorsed the necessity and the current system of reserving some scholarships for students from SC/ST backgrounds.

*'The Meritocratic Argument Against Reservations'*

Three of the ten scholars who opposed reservations strongly upheld the principle of 'meritocracy', i.e. the entitlement of only those who succeed in competitive examinations by their ability and effort. Shree (1964) put her case forward in terms of the 'disenfranchisement' of the meritorious as 'being against national interest'.

> I am… from a privileged background […] I went to some of the best schools and colleges and universities and we were top of the class. And I belong to a higher caste… Once you are not in [the] shoes of [SC/ST students], you can't speak for them, but you know, in the South also, all the Brahmins have been kicked out. When you are creating a quota, it is always at somebody else's cost. If there are fifty scholarships… and you keep ten as a quota, then ten others are being denied, who would have otherwise got in by merit. In trying to promote a quota, you are disenfranchising some other deserving people. Is that fair? Is that in the long term interest? Some of them may have had a better opportunity to promote science or talent or whatever you are promoting. … In the long run, this becomes a self-promoting bureaucracy. Nobody wants to give it up. Originally in India, this Scheduled Caste, Scheduled Tribe quotas … had a time limit in the Parliament also. 'Now they'll come up to power and the playing fields will be equal'. The playing field is never going to be quite as level. So I feel […] there are already quotas, every-where in India. [..] Maybe if they are trying to promote talent, they should invest in the lower end. At the prathamik (primary) and elementary education level. They should give them more money to come up at the beginning end…. Agar kuch equal karna hai[1], it is better to invest in the primary stage….Right at the primary school stage, at the secondary school stage, have some incentives for teachers to go and teach in a primary school, have some special initiatives in these underfunded,

rural, scheduled caste, scheduled tribe areas. Let us say they do this ... So that child, by the time they come up to the higher secondary, they are at par or almost at par with the so called privileged students. I don't think a talent exam should have quotas. .... I don't think in the long run, that would be serving our national interest.

All ten scholars who were against reservation echoed this point, i.e. talent search is the wrong level at which to intervene on behalf of such students. Rather, it is from the primary education stage onwards where the state should intervene so that by the time these children come to Classes XI or XII, they are able to participate on par with others in tests like the talent search. The interventions can take the form of 'special coaching, special provisions', i.e. 'facilitation' (Rupa, 1970), special 'upliftment' programmes 'so that the people can educate themselves' (Anil, 1980), 'similar opportunities for learning' as the better off (Roshni, 1970) and making 'sure that they get good schooling [because] if all go to the same kind of school, it won't be a problem' (Manas, 2000). Roshni (1970) summed it saying, 'Whatever disadvantage they have, it should be addressed, rather than diluting the test'.

The common aspect to all these recommendations was the scholars' absolute confidence in the intentions and ability of the state to intervene successfully to address this situation. The only person among these ten who expressed an implicit doubt about the capacities of the state was Punit (1984), who suggested that lessons could be drawn from how Indian cricket is managed.

I am not saying there needs to be some kind of quota system ... But I think [there should be] some way of fairness to that kid in Jharkhand. You know the kid who goes to some village school.... getting a low score in NTS ...[not being able to speak] English properly ... He probably would not be articulate and stuff that way. As a nation that's something that we have to think about. I mean, look at cricket. That must be the only privately run system in India which respects talent from a very early age and you can see the nature of the Indian cricket team change now. Dhoni is a good example, Sehwag is a good example and they are not typical 'English speaking – went to- Bombay-gymkhana – and- you know Dad would afford membership'- kind of a thing. So I think that's the question. Do we have the infrastructure at the grass root level to identify that? ...There is no doubt in my mind that the bell shaped curve for talent exists. My suggestion would be twofold, you know. One is to widen what we regard as talent. We do not see talent as being 'oh-you-can-solve-these-multiple-choice-questions-and-you-know-the-bookish-kind-of-a-thing' but can you think outside the box, can you think differently? And I think, [we must] widen opportunities for people who are more disadvantaged. That would be my recommendation.

*The 'Creamy Layer' Argument Against Reservations*

Another argument was that reservations do not reach the most deserving of a group – just the creamy layer, i.e. *'the more well off or influential groups'* (Anil, 1980). Apart from Anil, Mani (1977) also made this argument. The difference between them was that Mani was willing to consider reservations even within talent search programmes, provided that it was based on economic, rather than caste grounds.

> I don't see why [reservations] should be determined by ... what caste they belong to. There may be a Brahmin boy who comes from a poor background who should be supported as much as because he may not have access to resources. Everything should be driven by economic guidelines. In the case of United States for example, anybody can go to Harvard or Stanford or other top schools; the university completely covers them. It is not based on what their origin is; it is not based on what their status in life. Ok, they have their economic means but if they don't have the economic means and they are very bright, the university just supports them fully. In the NCERT scholarship, reservations should be based on... economic background.

*'Reservations Are Patronizing'*

A third argument was that interventions such as reservations would be considered as patronizing and insulting to those who are truly talented among these groups. Manas (2000) put it as follows:

> Well it depends on how you select students from those backgrounds. If you purposely make cut-offs very low and make things very lenient for them, then it really serves no purpose... To think that just because someone is from SC/ST, it will make them less talented is a very prejudicial thought. In many ways, it defeats the purpose of non-discrimination....

Usha (1975) echoed the same thought, comparing the situation to a hypothetical situation of reservations for women.

> ... I feel that sometimes... you ask people who are probably capable to underestimate their capabilities. I will be very snubbed if NTSE may say, "She is a girl and she needs to do only this much of a project and we will see and then we will take her". I don't want that....I am willing to do the biggest problem ...

*'The Social Justice Argument for Reservations'*

Of the five scholars who supported reservations in the talent search, two of them, James (1968) and Hari (1969), developed a case for reservations because of fundamental flaws within the social system. James (1968) identified a deep-rooted caste bias at all levels of education, as follows.

> Remember the bias is built into the system because the creators of the system are upper caste people and urban people. So they want their children to excel. They want their children to get through. I am accusing the creators of these exams.... that they are biased in favour of their own children, in favour of their own castes, in favour of their own communities, in favour of their own classes. They discriminate against farmers... against rural children... against the poor getting into this whole system. [Reservation] is a good attempt. That is why in the university system that you see, that is where reservation works best...For example, I [have been teaching] post graduate journalism since 1987. Some of my best students are from these reserved categories, who would have never had a chance to get into the university, had it not been for the reservation system. The open merit students [...] would have literally wiped them out.

Hari (1969) was more reflexive about his personal advantages and how they translated into 'talent'. In this context, he examined the role of 'quotas' as a way of promoting equity in a 'corrupt' society.

> Here I was – the son of 2 academics who went to a private school- an English speaking school and exposed to lots of books and so on. If I did the same as some tribal kid from a village, then I should say that the tribal kid from the village is more talented than I was. Because he can't have all those advantages and still did well, and so, I think that has to be taken into consideration. I don't know about rigid quotas. The problem is corruption. India is fundamentally a very corrupt society....That is why people go by exam results. If they leave it to individual judgments, then people will come under all sorts of pressures from local politicians, from rich people wanting their nephews getting admitted. .... If you have a more honest society, then you could leave this to the people who are doing all the selection in the committee. 'Wait a minute, this candidate is really good and he is from a village; he has not done quite as well as this guy who has scientist parents; the difference should have been much greater. So really this kid is probably better and we should give him a chance.' That will be the honest way to do it. So in the absence of that, they have quotas. I think maybe quotas are one solution.

## The 'Diversity for Excellence' Argument for Reservations

Dev (1977) argued that reservations are a good tool to bring diversity in the profile of the students selected in any competitive examination. He made the argument from the perspective of the cultivation of excellence.

There is no reason why reservations [should] not exist [in a talent search]....We must encourage excellence and we must expand the group from which we seek excellence. We have somehow very artificially restricted the pool from which we look for talent. In a country of 123 crore Indians, we keep looking for excellence in about 1 crore of Indians. Expand the pool, you will get lots and lots of talented people.

## Assessment of Various Aspects of the NTSE

A second set of queries required scholars to critically assess the NTSE as an intervention. The following themes were touched upon: a) the relevance of the NTSE's goal of identifying and nurturing talent to build the nation, b) the pros and cons of the testing strategy, c) the age of the selection of candidates, and d) an assessment of the benefits of the programme.

### *The Relevance of the NTSE's Goal*

The scholars were asked if a scheme like the NTSE, which was conceived to identify and support talented students who would then contribute to nation-building, was still relevant.

The majority of the scholars (18 of 19) admitted that the goal of identifying and nurturing talented students for the purpose of building the nation was still significant (Table 7.6). The significant chronological pattern is that post-1976, there is a drastic decrease in the confidence in an examination like the NTSE in achieving the goal. Additionally, more scholars after 1976 (3 as opposed to 2 in the previous decade) agreed that opening the scheme to disciplines other than the basic sciences was a positive development. We can further analyse these responses under two heads.

### *'The Value of Various Disciplines with Respect to Nation-building'*

The most articulate defence for why it is necessary to have a policy which systematically identifies and nurtures youngsters who have an aptitude for science was mounted by Hari (1969), himself a scientist, and Mani (1977), a technologist-turned scientist. These arguments echo the original purpose for which the National Science Talent Search Scheme was designed. Hari put it as follows:

There is no question that at the policy level that ... India needs more scientists... Without scientists, neither engineers nor doctors will understand the basis for their fields. ... Science is the fundamental basis for all these other applications... From that point of view ... you do need scientific research to advance knowledge... to apply new knowledge, to create new applications and that is how modern societies, advanced

*Table 7.6* Scholars' assessment of the goal of the NTSE

| Assessments | Total no. of scholars | Scholars between | | |
|---|---|---|---|---|
| | | *1964–1969* | *1970–1976* | *1977–2000* |
| The goal of identifying and nurturing talented students is still relevant | 18 | 7 | 6 | 5 |
| The provisions of the NTSE are inadequate to meet this goal | 9 | 4 | 4 | 1 |
| The scheme should have been restricted to the basic sciences | 3 | 1 | 2 | 0 |
| The opening of the scheme to disciplines other than basic sciences was a positive development | 5 | 2 | 0 | 3 |

societies, work. There is a debate whether a country becomes rich first and then has very good science which helps it to maintain its wealth to keep getting richer or [if] science can help you to make you to get rich... If you look at Japan or US, both of them became strong economically before they became very strong scientifically at the cutting edge. .... I am not sure that you can have cutting edge science because science takes money and takes infrastructure and all of that needs a long term vision. But once you become sufficiently good economically like Japan did, if you don't have science you would always be borrowing ideas from other countries –you always be behind.... So I think that is one thing. The other thing [is that] a [new] country may need scientists. There are many areas which may be ignored by the West. You may need research in malaria, tuberculosis or particular agricultural problems specific to a country. Who is going to do that research? The country needs its own scientists to do research in neglected areas important for that country. That is another reason why each country should produce its own scientific research establishment and not necessarily wait till it becomes economically strong.

However, Shree (1964) used several elements of the above argument to question why the NSTS was not opened to applied science in the early decades.

I always felt, why didn't they allow us to do applied science? I mean, there is a lot of research in science which can be done through applied science. Especially in technology related to engineering. But in those days they didn't. They themselves, obviously, they themselves realized that this notion was flawed ... about only restricting it to pure science. Physics, chemistry, biology, like that but not relating it to engineering or medicine or branches of applied natural science. But now you are saying that they have realized, they are scouting for talented young people, whether they decide to go for social sciences or languages or whatever. Everything is essential.

*'National Interest versus Self Interest'*

Of the scholars who endorsed the goal, there were many who interrogated what is meant by 'nation-building'. Consider, for example, the case of James (1968), who gave up a professional career in science for a more activist role, being involved in people's science movements. He raised questions on the nature of the relationship between science research and the building of the 'nation'.

> Many of the Science Talent Scholarship winners ... never had to do anything with people's science. They just remained in their institutions and most of them are now feeding multinational corporations. The nation is completely put aside. ... The nation is constituted of various [groups]...[such as] the ordinary people of India... the farmers of India... the women of India... the children of India. Scientists and technologists end up serving corporate science or government science. So this is a very complex problem and if we look at nation building and the scholarship, I can put huge questions on whether any of the scientists and technologists who got the scholarship and are doing excellent science today contributed at all to nation building.

Usha (1975) also self-reflexively developed the same theme, exploring the commitment to doing research in science as being primarily vested in furthering one's own professional trajectory.

> Nation-building is such a weird thing. I think that is a sort of catchy phrase [used] by these [people]. But do I build the nation? No, I didn't build the nation but I do a lot of work in general science. [I write about feminism] and if my critiques would have been taken it seriously (but they haven't been taken seriously), then [these critiques] would form an important contribution for the nation. Do I think people like S--- or Sh---- in TIFR[2] are building the nation? No, they are pursuing their own career goals and I don't know about that. I don't know when a scientist is working in a lab is pursuing just their own interests or [if] that is nation-building. I don't know what is meant by nation-building.

Both Malini (1975) and Manas (2000) explored the same question from the perspective of brain drain and, in fact, they defended the right of individuals in a democratic country to pursue their self-interest. Manas' argument is interesting because he embraces the perspective that one can legitimately ask 'what the country can do for you', rather than 'what one can do for the country'.

> I'll give you an example. IIT Kharagpur has this very impressive campus and it was the first IIT in India. In its main building, which is a thirteen storey building, there is a huge header which says, 'Dedicated to the service of the nation'. Those fonts are almost half a storey big. People used to wonder which nation we are dedicated to the service of, whether it is India or it is the US. (Laughter). Yeah... things [have] improved in India, so career wise, you have a lot of people coming back to India. Most of the faculty in our department are people who did their higher studies in the US or in Europe. And that is true for all the departments of IIT Delhi. So I think that has begun to change in a small manner.... in fact, a little while ago, I was having a conversation with another faculty [member]. We were having a discussion about the differences between India and the US. If you make it broad enough and less elitist, you will find that people who are from more disadvantaged backgrounds in India would rather not stay in India. Because they have only seen bad things happen to them in India. ... If that person decides no matter what scholarship I get, I want to set my foot in the US or in the UK or Europe or some developed country and never get back to India, I would not hold him or her responsible for that. So that pretty much is a conundrum.

Other scholars also argued that the programme must account for the aspects which draw individuals into a career in science or cause one to reject it (Anjana, Vinay, Usha, and Malini). Anjana (1964) provides an example of their position, which critiques the NSTS's understanding of how 'talented students' opt for a career in science and the general difficulty of scholarship schemes in incentivizing this trajectory. She also argued that at a personal level, 'nation-building' is the least of an individual's concerns. Individual's choices about their career are shaped by the social images of careers as well as closer sources of influence such as the family.

> I think today, one has to be hugely motivated and passionate about science to get into science because science is not a paying field. I see so many young people who look at their contemporaries who go into management and all, not only do they earn very good salaries but also get promoted fast. Science is a long drawn career. You do a Ph.D, then you do a post doc and then only you can think of being absorbed. Today an exam like the Science Talent is totally irrelevant. Science talent, in my time, it only reached some cities. The way, that the Department of Science and Technology, this INSPIRE programme, the way it reaches so many

youngsters and so many people, I have gone to an INSPIRE camp to give a lecture and I find that most of the students are there due to parental pressure, they want to do engineering or maybe medicine. But basic science is not an option which is considered by many bright students.

## Testing Strategy of the Talent Search

The scholars were asked to give their opinions on the testing strategy of the NTSE as well as to identify aspects which they felt needed to be revised (Table 7.7).

Of 18 scholars, only six felt that the test in the form which they had given it was effective in identifying talent. Five of them were winners between 1964 and 1970. Five other scholars held that the exam can be made effective with some changes which incorporate some aspects of its earlier NSTS form (though not returning to the science-only format). Rupa (1970) provides a typical instance of their argument:

> I wish they had retained project and interview. Because the interview tests the conceptual understanding of the scholar. And the project, I still remember the project I did and that activity helps. It shows the perseverance of a scholar, rather than just taking a paper pen test.

Eight scholars held that the test was ineffective in identifying talent. The critiques which they level are taken up in detail below.

### 'Bias towards Conventional Academic Success'

The first argument, made by six of these eight scholars, is that the test is biased towards certain types of intelligence or competencies which are required for conventional classroom academic success. What all these scholars hold in common is the critique that the test does not really look for 'out-of-the-box' thinking. Suma (1974) argued that successful classroom academic skills necessarily do not imply that an individual will be a creative scientist. The skills required for success in scientific pursuit may also develop

*Table 7.7* Scholars' assessment of the NTSE testing strategy

| Assessment | Total no. of scholars | scholars between | | |
|---|---|---|---|---|
| | | *1964–1969* | *1970–1976* | *1977–1985* |
| The test is effective as it is | 6 | 4 | 1 | 1 |
| The test is ineffectual | 8 | 2 | 3 | 3 |
| The test can be made effective with some changes | 5 | 1 | 2 | 2 |

later in life and also in ways which are not acknowledged by a test like the NTS.

> What you test here is [whether] the person is a good student and these are people who would have been good students in science,...not necessarily creative in this one thing but general good students...Some of [my very good students] have said that they were terrible students (in school).... I mean very average students at that level. If you have a talent test at class 12, many of these students would not have been selected. Some of them blossom after coming to [her institute] because they find out other students around them who are interested. So they learn from each other'.
>
> Suma (1974)

### 'Bias Towards Upper Caste and Class Students'

The arguments of four scholars also demonstrate an awareness that official interpretations of 'talent' are rooted in particular social, economic, and cultural contexts. In their reflections on the strategy through which 'talent' is identified in the NTS, they held that their own social position and the kind of family they came from gave them an advantage over others. Two scholars (1968, 1984) also made a sociological argument that the kind of talent that the test favours is biased towards the upper castes and that it is rooted in the divide between 'thinking' versus 'doing' or 'making' with one's hands. For example, James (1968) put it as follows:

> The whole idea of intelligence is upper caste biased. ... Nothing to do with kinaesthetic [intelligence]. What about crafts? Woodwork? Where is woodwork evaluated? Where is pottery evaluated? Nothing. The artisan classes simply do not stand a chance. The intelligences that are required to be a good carpenter or a goldsmith or a carpenter are not even considered in the science talent scholarship.

In addition to caste, Dev (1977) was forthright about the role that class played in creating dispositions for success in tests like the NTSE:

> If you look at life opportunities....you can see the very significant role of money. How much money.... A generation ago, it was your parents' educational background. But in the last 25 years, people have been able to compensate for that. Now cash... So a few things stand out- money, parental education, how much interest their parents' take in their child's education, what is your location- what kind of school you went to... If you give me these three variables: ... rural urban location, type of school and parental education and occupation, and I can make an intelligent guess, where that person's career would end. That is a telling indication of how much of a farce this thing called talent is.

*'Inadequacy of One Test for Different Aptitudes'*

Two scholars also questioned the effectiveness of using one talent search test in identifying aptitudes and skills in several areas. On the other hand, Anil (1980) summed up the argument of why having just one test for all the subjects defeats the purpose of having it in the first place.

> People, who are talented in one area, might not be talented in another area. I don't think…I have any talent in the social science area. But I bet a lot of people are good at that…I mean what is the objective of the scheme? If the objective is to identify talented students in different areas, then you need different schemes for each area. It is not that talent is universal. Most people who do well in one area might not do well in another area. So it has to be subject specific.

In all these eight cases, the scholars recommended an overhaul of the current testing model.

## The Age of Selection in the Talent Search

Scholars were asked to recommend what they thought to be an apt age for testing the talent of students. Their responses are tabulated in Table 7.8.

*'The Earlier, the Better'*

Krishnan (1970), Usha (1975), and Dev (1977) held that the earlier one identifies talent, the better. They make the same point in different ways, i.e. the earlier we identify students with potential, it is possible to socialize and educate them to channelize this potential effectively. The emphasis is on the kind of institutional support which is needed to make such a proposal effective and sustain it.

*Table 7.8* Scholars' assessments of the appropriate age of students in a talent search

| Recommendations for age of testing | Total no. of scholars | Scholars between | | |
|---|---|---|---|---|
| | | *1964–1969* | *1970–1975* | *1976–1984* |
| The earlier talent is identified, the better | 3 | 0 | 2 | 1 |
| The age group of 15–18 is best | 4 | 1 | 1 | 2 |
| 15–18 is too early for a child to make up his or her mind | 5 | 2 | 2 | 1 |
| There should be multiple points of testing, rather than one fixed age | 3 | 0 | 1 | 2 |

Dev (1977) pitches the argument that earlier identification of talented students is necessary for ensuring greater diversity in the pool of talent.

> And if for some reason you are not capturing people of a diverse social profile,… then the only way is to go early, to catch them young, to catch them younger and let their talent flourish….Do it at eight, nine, ten, identify their potential and then give them guarantee of passage through the school, eh, scholastic system.

As a bureaucrat, Krishnan explores the issue of age of selection and nurture with respect to the idea of individual transformation. The example that he chose to explain his point is the civil service.

> Yes, eight to ten will be [an] ok [age]. From the time they can be away from their families, bring them to an institution which will really foster [them]. Why does the IAS work? Inspite of very ordinary or below average people getting in, it is the system of selection [which works]. You are brainwashed to think that you are a superman, so you are forced to act like a superman and you become one. Some of the young kids who come [make me] think, 'How did he get in?' Then for three to five years, he is doing tremendously [well] and then in ten years, he [starts] to [decline] because the system is eating him up.

His use of the term 'foster' resonates with the description of talented students in educational policy documents as 'the wards of the state', where the state is presented as a foster parent. His example though presents this as Janus-faced situation. The state machinery which trains bureaucrats is presented as doing a transformative job. However, the warning is the lack of a long-term state perspective about individual development.

## 'The Later the Better'

The second category of responses was those who held that the end of secondary schooling is the apt period for the selection of the talented. These individuals were themselves selected at this age and saw no reason why this pattern should not continue.

The five individuals, who held that even the ages of 15–18 might be too early for the identification of talent, present several nuanced arguments for our consideration. Four of them explored this idea in terms of their own personal trajectories and what their selection at that age implied.

Shree (1964) had plans of pursuing medicine but the stipulation that NSTS winners must take up basic science led to her pursuing chemistry, a decision she regretted for several reasons. She found it very hard to find a job in the United States. In the area of medical research, her experience was that a PhD holder in Chemistry was less likely to get grants or lead research positions. Finally, she gave up her dreams of medical research and is now a professor at a Community College in the United States.

In the case of Hari (1969), he could not qualify for the IIT exams, and he could not pursue his desire to do medicine. Therefore, at a point of personal confusion, he took winning the NTS as a sort of *'message'* that he should pursue physics. Later on, he switched to chemistry and won the highest laurel in science research in the world for his area.

Punit (1984) did not feel compelled by the scholarship to pursue any particular line, but he took up engineering because he felt it was expected of him. He found the programme very stifling and restrictive for someone of his nature. It was only by his masters that he found in the area of 'design', a field where he could pursue his love for technology and creativity. Therefore, their reservations regarding the identification and streaming of individuals even in late adolescence are rooted in their own personal nego-tiation of their professional trajectories.

Three of them, i.e. Hari who is now based in the UK, and Shree as well as Punit, who are in the United States, elaborated on this idea with respect to the greater freedom that one possesses in these countries with regard to choosing and planning one's profession. Shree (1964) presented the Indian situation as one of 'entrapment'.

> But even at eighteen you may not know which direction your full potential is? You're too young and you discover things along the way. You may enrol into this, but at least in our system of education, you are in a narrow route at a very young age and you can't get out of that.

Hari (1969) uses the metaphor of 'entrapment' as well.

> Just because someone shows some aptitude when 11 years old doesn't mean what they want to do later on in life, right? I think the child should be exposed to a variety of subjects and not streamed into some direction. This is something Soviet Russia used to do. Pull out people who were athletic, good at chess or music and train them and coach them to get to very hard levels. Many people benefited but I bet some of them would not have been happy because they may have really wanted to do something else. I think people need to find themselves and this is a very decadent Western ideal. People need to grow up a little bit to find who they are and what they want to do before they get trapped'.

He reflexively uses the idea of 'freedom of self-exploration' being a luxury in a country like India, which has to deal with poverty and educational attrition.

Suma (1974) and Punit (1984) added the dimension of what 'not being chosen' at this stage implies for those who are not selected. Apart from the perspective of an educator, Punit also compared the differences between his son's and daughter's response to school and the institutional interpretation of their experience.

My son sort of figured out school pretty early and they do a lot of tracking. So if you're good at X, they put him in the advanced class and this that and the other. My daughter takes a little longer to get school. So that puts her at a disadvantage in the sense that it's not that she's any less smarter. It's just a differ kind of smart and schools are like just one kind of smart, which was, sort of, my problem in Pilani. You study something today, tomorrow there is a test, day after we study something else. There is no space to reflect or think about it and sort of get it in your head in the right way. See that's part of the whole multi-dimentional thing, I worry about. Identifying kids too early, not for the kids who get identified as being the special ones but for the other ones, who don't. Because they might be special too, just that they take longer for that to come out. You see, that's the issue. They see themselves as being deficient and that's a terrible place to put a child in. So the effect of NTS is not just on the kids who get it. It's also on the kids who don't get it. "Because that kid got it [and not me], it means I must be dumb".

However, Hari (1969) does take up this point and he ruminates on what such tracking might mean in the Indian context which has to deal with the reality of poverty and educational attrition.

I think you can look at it both ways .... The idea of pulling out bright kids giving them and giving them extra ... very good education is not a bad thing. I know a lot of liberals don't like it because it means other kids suffer because bright kids are gone and you are left with this school without bright kids –left over kids suffer and they don't like the idea of this track. So many liberals in education don't like this. But I also feel that if you are from a poor family, this may be your only way of getting a good education.

## 'Multiple Points of Entry in the Test'

In contrast to fixing one age at which the test can be taken, three scholars, i.e. Malini (1975), Mani (1977), and Dev (1977), suggested the use of a system of multiple entry points into the talent search programme. Dev put it as follows.

If for some reason, you are not capturing people of a diverse social profile, it's your mistake. Then the only way is to go early, to catch them young, to catch them younger and let their talent flourish. That is the only way. That is what I was sort of proposing – to bring [the age of entry] down, not to do it at plus two level [but] do it at eight, nine, ten, identify their potential and then give them guarantee of passage through the school, eh, scholastic system... upto Ph.D.

*'The Age Doesn't Matter'*

A different perspective was provided by one scholar who refused to commit to a specific age of identification. Making her argument from a disillusionment with educational institutions in India in general, Usha (1975) held that it is the lack of freedom that they experience in the conventional system of schooling that prevents the early manifestation of 'talent'. She shared her own frustration of being unable to 'fit' in the 'neat categories' that schools, college, and research institutions operate with.

> I don't think our schooling here is [oriented] at all towards the expression of getting the best out of children, identifying talent or anything. Our schooling is more of the stuff that you write your exams and that is the way our schooling moves... By and large, [it is] mass production, [children] going through [a] production line.

## Nurturance in the Programme

Scholars were asked to comment on the various aspects of nurture that they experienced through the NTSE and if they had any suggestions for change (Table 7.9).

### The Summer School Model

The summer school model was upheld as a suitable model for intervention by six scholars. These were all individuals who had attended the summer schools. The larger number of individuals who endorsed the summer schools (4) were winners from the first five years. For example, Arihan (1969) compared this model favourably in contrast to the practice in the United States of identifying, tracking, and separating 'gifted students'.

*Table 7.9* Scholars' assessment of the benefits of the programme

| Assessments | Total no. of scholars | Scholars between | | |
| --- | --- | --- | --- | --- |
| | | *1964–1969* | *1970–1976* | *1977–1985* |
| The summer school model is sufficient | 6 | 4 | 2 | NA |
| The summer school model is inadequate | 1 | 0 | 1 | NA |
| Need for revision of scholarship amount and number of scholarships | 1 | 0 | 0 | 1 |
| Other suggestions for change | 6 | 1 | 3 | 2 |

Gifted education [in the US]... I have qualms about gifts, ok (laughter). Tracking kids like that, I have a problem with that. It's a tough question, a tough question. How much do you do that? How much do you separate them from the rest? [In India], we never separated them. We were together. Our teachers managed... We [who were chosen in the NSTS] [were] not by ourselves, we [had] a surrounding, influencing others who are not there in the programme... [So if] I have seven other friends who are not in the Science Talent Programme... I am able to explain some things to them. Talk to them about my experiences or I can be an influence for them. This was something you were able to do when you were in school or college.

With reference to specific suggestions for change, the one scholar who commented that the summer school model was inadequate, Krishnan (1970), suggested the very alternative of the above. The reference to the quality of teachers is also influenced by his own disillusionment with the professors whom he had at the summer school.

I think it would have been better in those days not to [have] put us in the mainstream but make an institution like IIT.... Make an institution and take us all there. It sounds very elitist but that is the only way. Somebody at the top should see to it that people who come there as teachers are really superb.

Punit (1984) proposed that the content of the summer should be modified so the development of talent in one area should not preclude the possibility of other futures which the individual may choose for himself or herself.

If you've identified this talent in scholars, whoever they are, what you need to give them is a range of experiences that have to do with art and have to do with science and engineering because they are going to shift in life. You can't say just because you are good in maths, you are going to become a mathematician or you are going to become an engineer. You might be good in maths but you might end up being completely different.. a businessman... that's ok. I think this idea of talent needs to be opened up to possible futures that we create...That would be my take on it.

### Revision of Monetary Benefits and Number of Scholarships

Only one scholar, Dev (1977) suggested a revision in the number of scholarships and in the amount and he shared that he had placed this suggestion before the NCERT.

I was also proposing that the number of [scholars] should go up to some ten thousand. 10,000 fellowships, give 5000 rupees a month....

And strangely it was the NCERT which dragged its feet. They said, Itna paisa kyun dena chahiye? (Why should one give this much money?) And I said, Kishore Vigyaanik [Protsahan Yojana] mein itna dey rahi hain, toh yahan kyun nahin?. (If the Kishore Vigyannik Protsahan Yojana [a scheme for promoting youth to take up science careers] can give so much, then why not here?)

*Issues of Long-Term Economic Security*

With respect to the other suggestions for change, several scholars drew attention to the economic insecurity associated with a career in science and the lack of attention to this aspect within policy and programmes like the NSTS. For example, Vinay (1969) and Malini (1975) gave up a research career in science because of this aspect. Vinay described the influence of a professor at the Institute of Science, University of Bombay, in abandoning a research career.

> You know, I had a professor at M.Sc and he sort of discouraged that idea. He said unless you have something to fill your stomach, don't go in for it. First find a job, then settle down and then you can pursue your interest in science.

Usha (1975) pursued a research career in the basic sciences but described her regrets in this regard as follows.

> Science is a very unique thing. Being in research is very unique. Professionals – first when you get this money and you go into a profession – after that you are making tons of money; not one of them will go into rural service, not one of them does anything and even if it is forced on them, they will find ways of getting out of it and make tons of money. Whereas we guys never make tons of money. Even when we go abroad, compared to our peers abroad who went on to do medicine or corporate jobs whatever, we still make very low level of money. So a scientist s life is not loaded with money and actually I am advising people not to go into science. [Policymakers] have to smooth out the pathway and make things clear and [create] job security. They don't think about these issues. I have class mates here- my engineering class mates -who have taken voluntary retirement in 2004 and they are doing fun things in life…Here we are still struggling in some senses. It is very humiliating…So the structure of science – there need to be a rethinking.

Anil (1980), who is a rare example of an engineer who left a lucrative job after five years to complete his PhD and take up a research career in IIT, summed it evocatively using the relationship between 'talent' and 'glory'.

For channeling students, for channeling people to go into a research career, you should make the research career glorious enough. Ok… I mean, if in general…someone after doing research in the pure sciences is struggling to get a job and he is not able to support himself, that feeling percolates down and people don't consider that as a viable career option.

Therefore, a recurrent suggestion was that a programme of identifying and nurturing talented students should also be connected to a definite job placement.

## Discussion

The queries on the requisites for the development of an individual's talent as well as the factors that go into the non-recognition of talent allow us to explore the scholars' perspectives regarding the relationship between talent and identity.

### *Internal Self Knowledge versus External Recognition*

An important aspect of these narratives is the construction of individual identity, authenticity, and agency. What is evident is the assumption that the talented individual is ideally someone with a kind of self-knowledge about the possible futures which his or her ability can create.

In his essay, 'The Politics of Recognition' (2009), the philosopher Charles Taylor argues that this form of articulating the identity of an individual as something which is particular to them and which they discover in themselves emerged at the end of the eighteenth century. This ideal is related to that of authenticity, i.e. 'being true to myself and my own particular way of being' (Taylor, 2009, p. 28). Authenticity can be conceptualized as the basis of the massive subjective turn of modern culture. It shapes how one conceives the self as a human being with inner depths and with an original way of being human, i.e. one's own standard of what is possible in their lives. In the narratives of the scholars, we see the valorization of such a self-knowledge as giving the talented individual an agency with respect to his or her environment which those who lack this knowledge can't exert. While external recognition of ability is presented as important, it is the internal recognition of one's ability, which enables an individual to weather adverse social circumstances till they ultimately receive their due.

The problem with 'inwardly derived, personal, original identity' (Taylor, 2009) is that it does not enjoy an a priori recognition such as that which is derived from social contexts (birth, class, occupation, caste, etc.) which everyone takes for granted. Rather this recognition has to be won through a variety of methods and these attempts can fail. What has come about with the modern age is not the need for recognition but rather that the attempt to be recognized can fail (Taylor, 2009). This extent of one's dependence on society for the recognition of one's identity is also a modern phenomenon. The NTS scholars are not innocent of this understanding. A number of scholars presented the nuanced perspective that 'talent' is rooted in particular social, economic, and cultural contexts. In their responses to several queries such as those regarding the requisites for the development of 'talent' or the testing strategies employed in the NTSE, the majority of scholars held that their own social position and the kind of family they came from gave them an advantage over others. Their talent or potential had been recognized by their family or by mentors to whom they had access because of their social position.

What is significant here is that most scholars underplay the value of the recognition by the state through a programme like the NSTS, though they have experienced the socio-emotional benefits of its prestige. The state is presented as unable to understand their real requirements. However, this is represented as not affecting them because of the range of inner and external resources that these individuals have access to, by virtue of their privileged social context. For example, the tension between what the state considers as optimum for the development of an individual's talent and the scholars' own sense of the same emerges clearly in the discussion regarding the age of selection or induction into a talent search programme. The opinion held by several scholars was that even if a child is identified at a very early age, the absence of the requisite support system and a long-term perspective of his or her developmental needs render this moot. At the same time, very early identification and slotting of children into predetermined career trajectories stunt their agency in growing into possible other careers which they are interested in. The main critique of state schemes like the NTSE is that they tend to be very rigid and narrow in their conceptualizations of what forms talent will manifest itself in as well as regarding individual aspirations.

The contrast of this position vis-à-vis the state's role with respect to the talented emerges starkly however when the scholars were asked to comment on the low success rates of various population groups (girls, rural students, and SC/ST students) in the NTSE. In response, the majority of the scholars turned to the state as the provider of the means of redress in the cases of rural and SC/ST students. For instance, a popular suggestion was to equip rural students better through state interventions which focused on equipping their teachers. Another suggestion was that the standards of regional boards should be raised to allow these students to share the same kind of knowledge which students of elite boards like CBSE and ISC are exposed to. Similarly in the case of the SC/ST students, the state is presented as the agency with the power to ameliorate their situation. While the majority of scholars rejected the mechanism of reservations in competitive examinations as a suitable intervention, they argued that the state should intervene from the primary level of education onwards to ensure comparable quality and learning experiences for these students so that by the time they arrive at the stage of competitive exams, they are equipped to do so on their own 'merit'.

The discussion on the question of reservations in particular in many ways also reflects the fact that the sample of scholars who were interviewed for the study was predominantly from the upper castes and from minority groups like Christians, who have historically had greater social access to higher education. As many studies have demonstrated, this itself restricts the conversation around the question of caste, the very nature of the access to disciplines such as science and the implicit structuring of higher education institutions and their culture in modes that can be alienating to those from SC/ST backgrounds (Thomas, 2020).

The state is therefore presented as the alternative for the disenfranchised but not as the securer of welfare for the scholars themselves. The exception among the recommendations of the scholars for disenfranchised population groups was regarding women and their lack of success in a test like the NTSE. The notable thing in these narratives is that, by and large, scholars were able to present a nuanced picture of how socialization creates dispositions for success. None of the scholars attributed differences between girls and boys to a mere product of their biological make-up. However, the male scholars in particular didn't present any solutions to this situation, including the issue of special intervention by the state. Female scholars, on the other hand, critiqued the state of affairs from a perspective which privileged the self-knowledge and agency of the talented. They suggested that women need to exercise their own agency to come out of this social circumscription. They rejected any state intervention on their behalf as being 'patronizing' and 'insulting' to their sense of self as 'talented'.

### The Meritocratic Worldview of the Talented

Despite the reflexive character in how the majority of scholars interpret their ability in relation to their social advantages, they do manifest a legitimate sense of entitlement to their success because of their efforts. This is especially true of scholars who have pursued a career in basic science, engineering, or bureaucracy. Even in the analysis and recommendations of the scholars for groups which underperform in the NTSE, the emphasis is on the state providing the infrastructural support which, in the long term, will enable these individuals to compete on their own merit as the result of their own hard work. Therefore, what comes through as an essential aspect of the worldview of the talented (even in the above two themes) is its meritocratic character. A fundamental aspect of the meritocratic worldview is the tendency to 'locate the causes of events internally within attributes of individuals' (Kaiser & Major, 2006, p. 807).

One of the key indicators of this meritocratic worldview is that the majority of the scholars (except for two) did not engage with the structural foundations of inequality – foundations which may include their own practices. If the source of inequality is ultimately laid at the door of individual effort, then the professional work undertaken by the scholars can be constructed as having no 'role in that inequality nor... any responsibility to attempt to rectify it' (Cech, 2013, p. 75). Such a perspective being espoused by scientists and technologists is not the result of 'uncaring or naïve [individuals] but rather the outcome of a cultural frame that eliminates these complexities from problem definition and solution' (Cech, 2013, p. 75)

A similar point is made by Rukmini Bhaya Nair (1997) in her study of undergraduate IIT students as examples of the character-type bred under the conditions of modern technological education. She uses the students' responses to ideas about the nation, the role of technology, and the place of citizenship emerging from the writings of Nehru, Gandhi, Tagore, etc. to point to their dominant tendency to perceive themselves as producers of socially beneficial technologies, which then absolved them of obligations of social involvement or nationalist contributions. Their predominant perspective of India was that its poverty and culture hampered their personal and professional development. Nair locates such perspectives in their sense of the self which is shaped by technologies like the Internet and cybernetic education. These cause national affiliations to seem less relevant. In addition, their specialized training was seen as a means to pare away 'nationalist' attachments to India. Their elite education was a training in 'professionalism', which gave them valued status and opened 'limitless' possibilities, especially opportunities for migration abroad (Nair et al., 1997).

We see a similar line of argument being taken by the majority of scholars in the discussions of their role in nation-building. A number of scholars interrogated the very concept of 'nation-building', arguing that the commitment to doing excellent research in science for example is a commitment to further one's own professional trajectory. The point they raise is that at a personal level, 'nation-building' is the least of an individual's concerns. In their construction of 'nation-building' as a somewhat idealistic enterprise, the interesting point is how the scholars (except for one) were not able to link this to the amelioration of structural and social conditions which prevent the flourishing of women or Dalits or rural students, who are fellow citizens of India. Therefore, a predominant characteristic of the worldview of the 'talented' which emerges in the narratives of the majority of the scholars is its deep conservative character, which manifests no discontent with the status quo and which implicitly justifies the existing order.

## Notes

1 'If you are going to make something equal...'
2 Tata Institute of Fundamental Research, Mumbai

## References

Aerts, D., Apostel, L., De Moor, B., Hellemans, S., Maex, E., Van Belle, H., & Van Der Veken, J. (1994). *Worldviews, from Fragmentation Towards Integration.* VUBPRESS.

Cech, E. A. (2013). The (Mis)Framing of Social Justice: Why Ideologies of Depoliticization and Meritocracy Hinder Engineers' Ability to Think About Social Injustices. In J. Lucena (Ed.), *Engineering Education for Social Justice: Critical Explorations and Opportunities*. Springer.

Eidhamar, L. G. (2021). Dimensions of the Relationship between the Individual and Her Unique Worldview Construction. *Religions, 12*(3), 215. https://doi.org/10.3390/rel12030215

Kaiser, C. R., & Major, B. (2006). A Social Psychological Perspective on Perceiving and Reporting Discrimination. *Law and Social Inquiry, 31*(4), 801–830.

Nair, R., Bajaj, R., & Meattle, A. (1997). *Technobrat: Culture in a Cybernetic Classroom*. Harper Collins.

Taylor, C. (2009). *The Politics of Recognition. In Multiculturalism*. Princeton University Press.

Thomas, R. (2020). Brahmins as Scientists and Science as Brahmins' Calling: Caste in an Indian Scientific Research Institute. *Public Understanding of Science, 29*(3), 306–318. https://doi.org/10.1177/0963662520903690

Vidal, C. (2008). Wat is een wereldbeeld? (What is a worldview?) In H. Van Belle & J. Van der Veken (Eds.), *Nieuwheid denken. De wetenschappen en het creatieve aspect van de werkelijkheid*. Acco.

# 8 Seeking 'Talent', Finding the 'Nation'

The span of more than 50 years of a programme's evolution has been a considerably vast terrain to cover. We have traversed this history of the National Talent Search Examination (NTSE) and found our bearings, being guided by both the larger perspective of the Indian state's policies for the 'talented' and the evolving self-perceptions of those who have been labelled so by this programme. This journey has allowed us to glimpse the power which is entrenched in the trope of the 'nation'. Like the ever-changing perspective on an upward mountain trek, the relationship between the nation and its 'talented' students has been sometimes startlingly clear and at others, obscure. Having reached the summit of our exploration, so as to speak, now it remains to conclude with a survey of our findings.

The use of multiple methods has aided our access to this final panoramic perspective. This has included an extensive survey of different kinds of literature: official documents on the NTSE and its incubation under the National Council of Education Research and Training (NCERT) (including 5 specific reports on the Talent Search from 1963 to 1967), 50 annual reports of the NCERT (1963 to 2013), information from 19 states and 4 Union Territories on the conduct of the State Talent Search level (sourced through the Right to Information Act), key policy documents on education and literature on the scholars selected through this exam. This data has been contextualized against the narratives of 19 scholars who were beneficiaries of the programme between 1964 and 2000 and who were selected for interviews in this study through convenience sampling. Having analysed this vast body of material, I will first draw attention in this chapter to some aspects which stood out as key findings. These are then discussed in the subsequent section, where I explore the relationship between 'talent', 'excellence', and 'images of the nation'. This provides a background which helps us to locate how the talented are imagined in the enterprise of nation-building. The final sections of the chapter present the implications of the study and the possibilities for future research, before concluding.

DOI: 10.4324/9781003344902-8

## Main Themes

In this section, I discuss four key aspects which help us to make sense of the history of the development of the NTSE. Rather than take up the findings in a chapter-wise manner, I have sought to highlight the predominant themes in relation to the discourse of talent and the nation which have emerged through the analysis of the various types of data.

### Changing Conceptions of 'Talent' in the Talent Search Exam

The operational understanding of 'talent' within the National Science Talent Search (NSTS) was of a certain 'aptitude for science' responsible for a successful adult career in science research, which was presumed to exist as 'potential' in children. It was also assumed that this potential could be identified 'with precision' by designing a special test which conformed to 'modern standards of testing'.

However, the constraints and limitations within this conception of talent and its nurture in the NSTS were evident in several ways very early on into the scheme's development. For example, by the mid-1970s, it was evident that aspects of science talent search examination design such as the thought-type question, which tested 'scientific readiness', and factual questions, which tested the mastery of extra-curricular scientific content, were both biased towards the kinds of socialization and opportunities which students had received. Yet this insight remained marginal to the testing strategy. Additionally, the scheme's design of its nurturance programme was practically conceptualized within the ambit of science education as a limited intervention. It was meant to assuage to an extent the impact of issues such as the content of the curriculum, the pedagogy, and the infrastructural inadequacies which impeded the talented student's progress in science. However, NCERT's lack of engagement with the systemic and social factors which spur or impede the student's desire to pursue a career in the basic sciences obscured how success in the NSTS was shaped by factors such as one's economic background, caste, gender, rural/urban background, the kind of school one attended, and the medium of instruction, rather than the pure 'scientific aptitude' of candidates.

By the mid-1970s, these aspects were impossible to ignore, and the attrition of scholars from the scheme shaped its opening to scholars who wanted to pursue a career in engineering, medicine, or the social sciences, rather than just the basic sciences. In this major shift, what is notable is that the institution did not revisit its own assumptions on how talent was identified by the programme in the revised National Talent Search Scheme post-1977.

This lack of theoretical engagement with what was a major break within the discourse of 'talent' in the scheme significantly affected the impact of the scheme. An important implication was the pragmatic conception of the 'talented student' as one who would labelled so by virtue of merely passing the

NTSE. This redefined general perspective on 'talent', which was divorced from a perspective of its development according to the requirements of a specific discipline, was accompanied by a move away from a long-term planning for the 'nurture' of scholars. This is reflected in the deterioration of the nurturance programme, the decline of the monetary benefits, and the lapse of the book grant. These aspects demonstrate a disinterest on the part of the NCERT in thinking through the issues and conditions required for successful performance within any academic field, as was done in the first phase of the programme.

The conception of the talented student as merely related to his or her competency in test giving was reinforced by the growth of a coaching industry, which grew more robust post the mid-1970s. This changed NCERT's focus to modifying aspects of the test design to reduce the 'coaching effect' and to demonstrate the organization's commitment to objectivity and transparency in the process.

### Evolving Ideas of Nation-Building in the Scheme

In the first phase of the programme, the understanding of 'nation-building' in the scheme was to supply a future cadre of scientists who would take forward the design and implementation of the national developmental agenda. However, this theoretical objective did not translate to ground reality on the scale imaged by the designers of the NSTS. The revised NTSE scheme no longer manifested the earlier confidence that a 'talented scholar' would contribute to the project of nation-building in a specific way.

However, the idea of nation-building, which undergirds the second phase of the talent search programme, is a less publicized version. Consider the significant absence of the articulation of how selected scholars would contribute to the project of nation-building once they were allowed to pursue higher education in disciplines like the social sciences, medicine, and engineering, in addition to the basic sciences. This absence shifts the onus of 'nation-building' away from the scholar to the institution of the NCERT. In doing so, one sees the mundane, procedural strategies through which state institutions continuously sustain a certain national consciousness. The procedures of the talent search, especially those that define the ambit of the search with respect to the territory and to the democratic values of the nation, are instrumental in this process. For example, measures like decentralization of the exam and the creation of state quotas for the national-level test use the federal model of the Indian union (the centre–state relationship) in the relationship between NCERT and the liaison organizations at the state level. Features such as the reservation of a number of scholarships for students from SC and ST backgrounds or to those who are differently abled acknowledge the democratic values of inclusion and equity.

This institutional sense of the 'national' in the talent search, which is defined through the notion of maximum 'coverage' is also easier to cross-examine. The data presented in the fourth chapter raises questions on the extent to which one can consider the talent search to be truly 'national' and whether the winners represent a 'national' cohort in their diversity. For example, the division of the NTSE into a state-level and a national-level examination ostensibly seems to have created a more 'national' (in the sense of representative) distribution of winners. However, when the data is disaggregated at the state and district levels, we see that there are huge skews.

Post the establishment of state quotas for the number of winners sent at the national-level test, Maharashtra emerged as the major producer of winners between 1985 and 2004, with nearly twice as many winners as the next best-performing state, Bihar. Even within states, the distribution of winners does not always correspond to a uniform state-wise distribution. For example, only 14 states have at least 50% of their constituent districts producing at least one winner between 2001 and 2005. Twenty districts of the country (topped by Bengaluru Urban) have produced 40.2% of the total number of winners during this period. There is also a consistent under-representation of the number of girls among the winners. More than 75% of the winners each year continue to be boys. Similarly, most of the winners tend to be from upper-caste backgrounds. Despite the reservation of 175 seats for SC candidates and 75 for ST, those who are winners of the scholarships tend to be concentrated in a few states. Maharashtra produced nearly three times the number of SC and ST winners as compared to the next best-performing state, i.e. West Bengal.

In terms of participation, we see that less than 5% of the total number of students who were enrolled in Class VIII wrote the state-level Talent Search Examination in 13 of 15 states in 2010 and in 14 of 16 states in 2011. Fewer than 50% of participants in the state-level exams of 2011 to 2013 in six out of ten states had rural backgrounds. Even in the states with a significant number of rural participants, the problem is that these figures do not translate into the numbers of winners produced, who overwhelmingly tend to be from urban backgrounds. The kind of school [government (elite or regular) or private] and its location are the other significant determinants of whether a student has the information necessary as well as the confidence to participate in a test of this nature. The absence of engagement of the test with the realities at the school level has prevented NCERT from understanding the various socio-cultural conditions which may impede learning, especially in the cases of girls and individuals from SC/ST backgrounds.

### Impact on the Academic Trajectories of Scholars

The first phase of the programme as the National Science Talent Search Examination (NSTSE) had a significant impact on shaping the career choices of those scholars who participated in its unique undergraduate nurturance programme of month-long, discipline-based, residential summer schools. Between 1964 and 1976, the emphasis of the National Science Talent Search Scheme was to retain students with an aptitude for science in research careers. Since the careers of scholars have not been tracked by the NCERT, there are no total numbers available.

However, the data analysed in Chapter 6 suggests that the scheme did meet its objectives for a number of students, who are now engaged in research as successful scientists and academicians at the national and global levels. Nineteen scholars (11 men and 8 women), who were beneficiaries of the programme between 1964 and 2000, were interviewed as part of this study. At the time when they appeared for the talent search examination, the profile of these individuals was urban and predominantly middle class, with the major occupation of parents being academics, educators, and government servants. Only four of them spoke of financial difficulties at home during their days of schooling and college. 12 were from upper-caste backgrounds, 1 from an OBC, and 3 were Christians (3 did not disclose their background). 13 had taken up research careers (11 in the sciences and 2 in the social sciences). Eleven of the 13 scholars were part of the Science Talent Search Phase (1964 to 1976) reminiscenced on the value of the NSTS's undergraduate nurturance programme in creating an enduring interest in scientific research, whether they ultimately pursued a research career or not. Nine of these 13 did do so. They affirmed the importance of the summer school programme in helping them to be connected to 'like-minded peers'. A related finding was that its special impact on the six women who decided to pursue research in science. For these women, the peer group they encountered in the month-long summer schools provided them with an unprecedented intellectual challenge, which they had not experienced before during their schooling or even from supportive families. This spurred greater confidence in their academic ambitions and led them to explore unconventional careers in research.

Post-1976, the distribution of career choices of winners post-1977 reflects that scholars were no longer considering the provisions of the scheme with respect to planning their educational trajectories. A major reason for this lack of impact of the NTSE is the design of the scheme, which did not either emphasize a research career as its core objective or even provide opportunities for these scholars to be absorbed in a professional capacity after the completion of their studies.

However, what does remain is the prestige which has accrued to the scholarship and its impact on the scholars as well as in the public imagination. The prestige attributed to the scholarship varied depending on whether the scholar pursued a research career or not and on the period during which the scholarship was availed. Those from the NSTS phase who did not pursue a career in science tended to highlight the prestige of the examination and the ongoing credibility which this label continues to confer. For those scholars from the NSTS and NTS phase who were presently in academia, the award did not constitute a source of the validation of their ability beyond the college level. However, scholars who were part of the programme after 1976 acknowledged the social capital associated with the scholarship, a capital that filtered through its association with elite science careers, its claim of filtering the best students from the national pool, and the sheer longevity of the scheme.

### Perspectives on Talent Development: The State versus the Scholar

This history of the evolution and implementation of the talent search cannot be understood without contextualizing it within the larger framework of the state's discourse on talent. Chapter 3 had identified two ideas of talent that emerge in the explicit discourse of Indian policy, when mentions of the talented student and his or her nurture within the education system are raised. The first and predominant perspective in the policy documents is that 'talent' is a special potential for excellence which is present in some students. The second idea of talent in the policy literature presents the concept as the general competencies and abilities which may be possessed by all learners. In comparison to the first, this is not a prominent perspective and was explicitly articulated only by four policy documents.

These draw on two contrasting images of the nation in policy. In the first case, talent is linked to an image of the nation as a meritocracy, where both the assignation of individuals to positions and the distribution of socially valued goods and other rewards is based on one's potential and performance. Such social organization is believed to be the most optimum for the full development of an individual's potential and his or her contribution to the welfare of society. In contrast, in the case of the second meaning of talent, the underlying image of the nation is democratic. Here, every individual's potential for meaningful social contribution is understood in terms beyond his or her current performance. Rather, individuals are contextualized in terms of their histories and social backgrounds, so that factors like caste, class, and gender are understood to be crucial in shaping the ambit of their abilities and opportunities.

The implicit discourse of talent on the other hand draws attention to the conception of the talented student as a citizen, whose potential is available for the use of the nation's interests. This finding drew upon a close comparison of the Education Commission Reports (ECR) (i.e. Kothari Commission) of 1966 and the National Knowledge Commission Reports (NKCR) of 2009, two sets of policy documents that contain the largest number of references to 'talent' in post-Independence education policy. The discourse of 'talent' in the implicit policy language of the former represents the talented individual as extremely valuable and available for the initiatives of nation-building. At the same time, his or her potential is extremely vulnerable to being lost or wasted if not given proper attention. He or she is represented as passive, awaiting the trigger of state intervention for the development of his or her potential. In contrast, by the 2000s, especially in the NKCR, one sees a shift in the depiction of the 'talented'. There is now a greater acknowledgment of the agency of the talented individual as someone with a strong sense of what is desirable for their own personal development. For example, this is reflected in a greater consideration of what kinds of environment 'draw', 'attract', and 'retain' the talented. The changes in the implicit discourse of 'talent' at the level of policy reflect a gradual and larger social recognition of the importance of individual identity and choices in shaping the discourse of the nation, rather than the other way around. In other words, one sees a greater recognition by the time of the NKCR that personal potential and entitlement might be experienced individually in ways which might be significantly different from the state's perspective as articulated in official policy and programmatic structures. These differences shape choices and responses to opportunities, a factor which policy must now increasingly take into account.

It is precisely with regard to this idea of self-knowledge and agency that the narratives of the scholars who availed the NTSE between 1964 and 2004 differ from that of the predominant state perspective. In their narratives, the talented individual is ideally someone with a kind of self-knowledge about the possible futures which his or her ability can create. As Chapter 7 demonstrates, for the scholars between 1964 and 1976, the choice to avail the NSTS scholarship was driven by motives which are different from the state's assessment of the importance of the basic sciences. The provisions and incentives of the NTSE made available a certain occupational conception associated with the basic sciences, which was in line with the scholars' self-concept. In other words, the incentives of the programme mirrored what the scholars and society both deemed them to be 'worth'. The specific impact which the NTSE had for scholars between 1964 and 1970 is how it incentivized a career in the basic sciences as a desirable option in relation to others such as medicine or engineering.

A number of scholars presented the nuanced perspective that their self-knowledge as talented individuals was the result of privileged social, economic, and cultural contexts. What is significant here is that they underplay the value of the recognition by the state through a programme like the NSTS, though they have experienced the socio-emotional benefits of its prestige. The state is presented as unable to understand their real requirements. However, this is represented as not affecting them because of the range of inner and external resources that these individuals have access to, by virtue of their contexts of privilege. However, in contrast to the position they took with respect to their self, the majority of the scholars turned to the state as the provider of the means of redressal in the cases of the rural and SC/ST students who traditionally have had low success rates in the NTSE. The state is therefore presented as the alternative for the disenfranchised but not as the securer of welfare for the scholars themselves.

The exception among the recommendations of the scholars was with respect to the case of girls who also tend to underperform in the exam. Though the scholars were able to present a nuanced picture of how socialization creates dispositions for success, what is notable is that in redressing this situation, the issue of special intervention by the state did not come up. Female scholars, on the other hand, critiqued the state of affairs from a perspective which privileged the self-knowledge and agency of the talented.

They suggested that women need to exercise their own agency to come out of this social circumscription. They rejected any state intervention on their behalf as being 'patronizing' and 'insulting' to their sense of self as 'talented'.

Therefore, what comes through as an essential aspect of the worldview of the talented is its meritocratic character, where the causes of events tend to be located internally within the attributes of individuals (Kaiser & Major, 2006). Even in the analysis and recommendations of the scholars for groups which underperform in the NTSE, the emphasis is on the state providing the infrastructural support which, in the long term, will enable these individuals to compete on their own merit as the result of their own hard work. In this context, the majority of the scholars (except for two) did not engage with the structural foundations of inequality – foundations which may include their own practices.

It is possible to correlate such an attitude with the predominant tendency of the scholars to be sceptical of the enterprise of 'nation-building' as an idealistic project, which had little relation to their personal priority of advancing their professional careers through excellent work. Except for one scholar, none of the others were able to make a connection between 'nation-building' and the need to ameliorate structural and social conditions which prevent the flourishing of women or Dalits or rural students, who are fellow citizens of India. Therefore, a predominant characteristic of the worldview of the 'talented', which emerges in the narratives of the majority of the scholars, is its conservative quality, which manifests no discontent with the status quo and which implicitly justifies the existing order.

## The Power of the Ideal of 'Talent'

This book, in no way, imagines itself as an exhaustive representation of ideas of 'talent' in the Indian context. This is a small attempt to chart a trail through which one may explore how the roots of our common assumptions about talent influence our practical endeavours to identify and nurture this quality.

While there are considerable differences in how the concept of talent is understood in day-to-day life by individuals as compared to its official and institutional interpretations, there does seem to be a general consensus that this 'quality' is rare, desirable, and worthy of society's care and investment. One way to understand the importance attached to the quality of 'talent' is to consider its relationship with the ideal of 'excellence'. In the context of this study which has oriented itself towards the concept of 'talent' as one that is socially constructed, I do want to acknowledge that the power of constructs lies in how they are received and internalized by individuals.

As human beings, there seems to be an innate pull towards what we consider to be the 'excellent'. There is an emotional quality to our response to a highly novel solution of a scientific problem, an evocative musical performance, the display of physical prowess in a sporting event, the powerful logic of a particular style of reasoning, etc. To think through this aspect of the power of excellence in arousing such powerful emotions of attraction, I have found Edmund Burke's discussion of the sublime in his essay *A Philosophical Enquiry Into the Origins of Our Ideas of the Sublime and Beautiful* (published in 1757) to be useful. He argues that when one encounters the 'sublime', one does not respond to it by means of one's reason, but rather through passion.

> In this case, the mind is so entirely filled with its object, that it cannot entertain any other, nor by consequence, reason on that object which employs it. Hence arises the great power of the sublime, that, far from being produced by them, it anticipates our reasonings and hurries us on by an irresistible force. Astonishment... is the effect of the sublime in its highest degree; the inferior effects are admiration, reverence and respect.
>
> (Burke, 2001)

A radical insight which Burke provides is that the response to the 'sublime' is not just psychological or emotional. Rather, it is physiological and rooted in the response of all five senses to experiences which cause us to feel that we are in the presence of something which is greater than our self. What we might take away from this for our purpose is how every individual's idea of excellence is rooted in prior experiences of encountering something 'greater' before which one's self feels inadequate.

The social recognition of this emotional power of certain types of excellence influences perspectives of 'talent' and demands that individuals with the potential for excellence be recognized and honoured in society. This is what confers positions of social leadership upon the talented across cultures. Whether this leadership was concomitant with political power was, of course, dependent upon the area in which an individual demonstrated prowess. This is true whether one considers Tang dynasty China, where children under ten years who could read certain prized texts like the Analects of Confucius were given special positions and certificates or if we examine the veneration of the possessors of 'pratibhā' in the texts of classical Sanskrit philosophy as the moral leaders of an age. The same holds true in eighteenth-century French and American republican discourses, which argued that the possession of 'talents' was an endorsement of certain individuals according to 'natural law' for positions of leadership.

The focus of this book's engagement has been how there was a particular vision of leadership invested in the 'talented student' in the historical context of India during the 1950s and 1960s. The prestige of scientists and technologists had exponentially risen post the Second World War because the Allied victory was often interpreted as resulting from their superiority in war machinery, including the power of the atom bomb. In the case of science and mathematics, in particular, the popular perception was that success in these endeavours required innate potencies, more than training ('talents', 'aptitudes', 'gifts', 'abilities', etc.). Post-independence development was understood as economic growth driven by capital-intensive industrialization. This led to a valuation of knowledge and expertise related to science and technology, possessed by scientists and engineers who would lead the processes of agricultural and industrial modernization, comparable to the Western context. As India's first Prime Minister, Jawaharlal Nehru's own personal convictions about the role of science in national development shaped the political influence which was enjoyed by several scientists (such as Meghnad Saha, Shanti Swarup Bhatnagar, Homi Bhabhi, etc.) during that period.

However, along with the influence that scientists enjoyed, there was also considerable anxiety whether such scientists and professionals could become modern equivalents of a new Brahminical class. The question of whether conducting a national talent search was going to create an elite class through a rationalized process of testing had also been a real one in the minds of the designers of the NTSE. The early reports of the NSTS specifically include concerns that the talented students could emerge as a separate social group, if they were separated from the mainstream student population. It is true that during the first ten years of the NSTS, the month-long residential programme which recurred thrice during their undergraduate days offered the scholars the chance to form strong friendships based on shared interests. This is attested to by all the scholars (save one) between 1964 and 1976, who were interviewed for this study. However, once the month-long nurturance programmes were dissolved (ten years into the talent search scheme), there was no longer a sense of community and group identity between the students. What this state-sponsored process of searching for talent did was to rationalize and consolidate the social advantages of the privileged. The winners of the NTSE all attended elite colleges and became part of national and international academia or corporations and state bureaucracy. Therefore, while the NTSE did not create a new elite, these individuals, by virtue of their academic and professional associations, were and are able to exert significant influence over the policy decisions of corporations and governments.

However, the relationship between the discourse of 'talent' and its interpretation by the state in relation to the task of nation-building is not a hegemonic one. As discussed before, the state's discourse of 'talent' is dependent on a conception of the individual as a citizen, whose potential is available for the use of the nation's interests. This interpretation is challenged by the predominant perspective that we observe in the narrative of the NTS scholars, i.e. the relationship of the modern ideology of individualism and the 'ideal of authenticity'. Being true to one's self concept implies being true to one's own originality, which is something only the individual can articulate and discover. In this process, one defines oneself and realizes an inner potential, which is one's own. This implicit assumption underlies the ideal of authenticity and its concomitant goals of self-fulfilment and self-realization (Taylor, 2009).

The implication of this is that notions of advantage and disadvantage as well as a sense of personal potential and entitlement are experienced individually in ways which are significantly different from the state's perspective as articulated in official policy and programmatic structures. These differences shape choices and responses to opportunities. For example, the rejection of a potential career in basic sciences by an overwhelming number of selected scholars, especially from the 1970s, was in a great part because opportunities for an aspirational middle class had expanded, leading to a greater differentiation of career choices. On one hand, they had access to the benefits of a very subsidized and government-sponsored higher education in fields, such as engineering and medicine. But on the other, this was coupled with a growing disillusionment regarding professional opportunities and recognition in the country. The kind of high-paying and commensurably prestigious institutions which could accommodate these individuals were few. This led to substantial migration of professionals outside the country. In other words, the nation's need for scientists was not a strong enough motivation in comparison with opportunities for personal and professional growth, which could be available outside the country. Therefore, gradually, there has been a growth in scepticism about the extent to which investing in the special education of the 'talented' yields returns for society. We find this tension for example in the ways in which the IITs are both the locus of tremendous social desire and social disillusionment because its graduates are seen to be more preoccupied with using the gains of their high-quality and highly subsidized education to further their own careers.

What lessons can we draw from this historical trajectory of a particular relationship between the talented and the aspirations of nation-building? If we bring these concerns to the field of education, they make us reflect on the aims of education, i.e. 'our concepts of excellence in life and society for the improvement of which we want to use education' (Naik, 1982). These aims affect the content and process through which young children are prepared for life in society. As mentioned before, there are two dominant, recurrent, and contrasting ideas of 'talent' prevalent in Indian educational policies and programmes. On the one hand, 'talent' is understood as a potential for high achievement in some areas which is present in a few children. On the other, 'talent' is also used to a general potential for successful performances which is present in all individuals in varying degrees. However, in both cases, talent is fundamentally associated with individual potential and the conditions which are required for human flourishing.

However, the focus on 'individual flourishing' alone hides the potential of another level at which the tension between the identities of the talented individual and the citizen can exist. Bertrand Russell captures this tension as follows:

> Citizens as conceived by governments are persons who admire the status quo and are prepared to exert themselves for its preservation. Oddly enough, while all governments aim at producing men of this type to the

exclusion of all others types, their heroes in the past are exactly the sort that they aim at preventing in the present.

(Russell, 2010, p. 4)

By virtue of their abilities as well as opportunities, ideally, the talented should be 'critical thinkers' – those who will reflect, analyse, and deconstruct existing structures and who will provide new energy as well as vision to revitalize society. This is a conception of the talented individual which has no place in the state-driven understanding of 'nation-building'. Surprisingly, among the cohort of the NTS scholars who were interviewed for this study, the predominant perspective among the majority was also a socially conservative one, especially with regard to their analysis of inequality among various groups.

At the heart of the idea of 'quality' in the education that we provide for students is 'what qualifies as a worthwhile educational aim or experience' (Kumar, 2005). In this context, the perspective on talent that we endorse as a society reflects the kind of 'nation' which we hope to build or to actualize, and this in turn must affect the way in which the 'talented' are prepared to contribute to society.

## Lessons for Policy

In the Indian context, as I have mentioned before, the discourse of 'talent' is a marginal discourse because of the tension between the principles of democracy and meritocracy. However, whatever references to the desirability of identifying and nurturing talent exist, they are couched in terms which are general and vague. These suggestions for programmes (such as talent searches and special scholarships) tend to be piecemeal or tokens attesting to the country's endorsement of the idea of excellence. They also tend to be articulated as separate from the philosophy, objectives, and recommendations for the education of the rest of the children of the country. Therefore, interventions for the students who are 'talented' currently exist on the fringes of the larger education system. Nevertheless, there is a wide variety of programmes for the 'talented', which are very diverse and range from the provision of financial assistance for tuition, books, supplies and other materials, opportunities for internships and travel, mentorship programmes, and government-sponsored and subsidized coaching classes aiming to prepare able students from disadvantaged backgrounds, etc. They are also under the purview of different ministries, including the Ministry of Human Resource Development, the Ministry of Culture, the Department of Science and Technology, and the Ministry of Youth Affairs and Sports. However, these initiatives need to be contextualized within the respect to the global shifts in terms of engagement with concepts of intelligence, creativity, aptitude, and talent, which have occurred in the twentieth century. There is now greater attention given to the very constructs of intelligence and other markers of ability beyond the model of psychometric test measures like the intelligence quotient. Studies also challenge the distribution of ability along a normal distribution curve as well as draw attention to how the relationship between the individual learner and his or her environment leads to an ever-changing range. Therefore, a major implication of this study is the need for a systematic assessment of government initiatives for students labelled as talented in various fields and their impact.

Secondly, as we have seen, this is a period where the discourse of 'talent' – its forms, the means to identify individuals with the potential for success and models of nurturance – is attracting greater attention at the global and national levels than ever before. In the Indian context, one problem in this respect is that this discourse is more popular at the level of higher education and in the absorption of college students in jobs. A major reason for the inadequacy of such efforts is the lack of integration of such efforts within the mainstream of school education. This has been bolstered by fears that the special acknowledgement of students who demonstrate exceptional capability in different areas is tantamount to supporting elitism and a different standard of the quality of education provision for some. However, as this study demonstrates, there exist parallel interventions for talented students outside the mainstream and in the long run, the separation of the two systems leads to the failure of the objective of identifying and nurturing talent in all students.

A global shift within the field of education since the 1980s and 1990s has been a reconceptualization of the learner along the lines of constructivist psychology coupled with a greater thrust towards inclusion. For example, UNESCO World Conference on Special Needs Education: Access and Quality, at Salamanca, Spain, in 1994 highlighted that 'all children's educational needs can be satisfied more or less, within the regular classroom in a mixed ability setting'. Therefore, there is a greater global emphasis on not just paying attention to a few talented students but in helping greater numbers of children become more creative by assisting them in becoming independent and searching for new ideas (Gallagher, 2000). There is a growing attention to the category of children with special needs even in the Indian context. However, this category only includes children with learning disabilities or a variety of physical or psychological disorders. Those who demonstrate high ability in various areas don't find a place or provision in comparison to the kinds of efforts which are increasingly being adopted for other special needs students.

Such an inclusive perspective places great demands upon the teacher. However, even in the National Curriculum Framework for Teacher Education, 2009, the segment on special education does not acknowledge the need to prepare teachers to support students who demonstrate exceptional ability. Perhaps one reason for this omission is the entrenched perspective that the identification of 'talent' or 'giftedness" belongs to the domain of 'experts' in the fields of psychometry and psychology. Even within the NTSE, a major drawback of the scheme has been the marginalization of the teacher in the process of identifying students with high ability in various disciplines. This is what sets apart successful talent search programmes in other countries, including the Westinghouse/Intel Science Talent Search Programme in the United States on which the NTSE itself was based.

In the light of these observations, it is also important to assess the impact and potential of the NTSE in its present form. As the study has demonstrated, the programme falls far short of its objective of the identification of talent and the nurture of the selected students. While the prestige of the programme and its certification of some students does confer socio-emotional advantages for the winner, that alone cannot be the raison d'etre of its existence.

The NCERT needs to explore other models of identifying talent, considering the research findings which have demonstrated how the aptitude test discriminates against students from rural and disadvantaged backgrounds as well as accommodates targeted coaching to 'crack the examination'. There has to be a greater attention paid to the modes through which passion and interest in different fields can be expressed by a student, beyond his or her performance in a generalized aptitude test. The project component in the NSTS phase of the programme was one such good indicator for the sciences. When it was expanded to include the social sciences, medicine, and engineering, the elimination of the project component was ill-advised. The focus on creating a highly bureaucratized and objective test (with easy accountability to queries under the Right to Information Act) led to the elimination of the interview. When the modality of the test overpowers its purpose, then the programme fails to meet its objectives.

A fruitful comparison can be made with another talent search which exists at present in the country, i.e. Department of Science and Technology's programme called Innovation in Science Pursuit for Inspired Research (INSPIRE). As the INSPIRE website[1] declares, 'A striking feature of the programme is that it does not believe in conducting competitive exams for identification of talent at any level. It believes in and relies on the efficacy of the existing educational structure for identification of talent'. A multipronged scheme with three levels (for school students, college students, and doctoral/ post-doctoral fellows), the school education component is significant for its attempt to provide 200,000 students every year with a chance to develop and exhibit an innovative project. In its component for school students of Classes VI to X, the Scheme for Early Attraction of Talent (SEATS), The onus of the selection lies upon the Principal/ Head of the school. Each school is supposed to nominate one student each from Classes VI to X. Based on the nominations, the DST selects two students (one from Classes VI to VIII and one from IX to X) who are then given an award of Rs. 5000 each to prepare and present a project at the district level. Following this, a select number of students from each district present their work at the state level. This leads to the further selection of five students, who will demonstrate their project at the national level. At the national level, the five best projects are selected. Annual summer camps are also organized for students of Class XI, allowing around 50,000 students under this scheme to participate.

At the level of undergraduate education, INSPIRE offers 10,000 students a scholarship of Rs. 80,000 each for undertaking Bachelor's and Master's level education in the natural and basic sciences. No separate test is conducted. It offers the scholarship to a certain percentage of the top-ranked students in 12th standard Board Examinations, the Joint Entrance Examination (JEE), and the National Eligibility cum Entrance Test (NEET). In addition, winners of the NTSE, the Kishore Vaigyanik Protsahan Yojana (KVPY), the Jagdis Bose National Science Talent Search (JBNSTS), and Science Olympiad Medallists are also eligible for the scholarship. Students admitted to the Indian Institute of Science Education and Research (IISER), National Institute of Science Education and Research (NISER), and Department of Atomic Energy Centre for Basic Sciences (DAE-CBS) at the University of Mumbai are also eligible. The INSPIRE doctoral/post-doctoral fellowship programme also takes care of a concern which animated several NTSE scholars, i.e. the absence of assured research opportunities after one's Master's level.

What is interesting in this model is how INSPIRE does not waste its energies on developing the perfect test but plays to the strengths of the participating organizations in developing strong nurturance programmes for the selected scholars. Similarly, one of the reasons why the NSTS programme's nurturance component was so effective was the active involvement of university professors and practising scientists in the design of the programme.

Therefore, the NTSE should be reoriented to reflect the NCERT's strength in terms of school education, rather than just identifying students towards the end of their secondary schooling and designing nurturance programmes for college students. As I mentioned before, under the paradigm of inclusive education, now there is a greater focus on developing classroom strategies that can accommodate both children with learning disabilities as well as high abilities. Nevertheless, in terms of curricular attention to the needs of children with high capacities in certain subjects, there has been less attention in the Indian context. Therefore, there is a need to explore how such children can be accommodated, whether through the chance to attend advanced classes, after-school opportunities, opportunities for mentorship by experts in the field, and so on. There is also a need for the concomitant availability of syllabi and textbooks, which allow insights for advanced-level pursuits of topics as well as guiding reading lists. This material needs to be available at the SCERT/DIET levels and in school libraries. Considering its scale of reach, the NCERT is poised with the capacity to set in place a structure that allows for the connection of the district, state, and national levels to provide opportunities for such children.

## Paths Not Taken

This study has only scratched the surface of possible explorations of the social construction of talent. Even just taking the history of the NTSE, which has produced more than 30,000 scholars since its inception, one sees the pressing demand to follow up and explore the trajectories of these individuals. Only then can one provide a holistic assessment of the impact of this scheme.

Each of the levels (the formulation of policy, the NCERT as an apex organization responsible for the design and implementation of education policy, and the perspectives of the beneficiaries and stakeholders of a programme like the NTSE) merit further investigation, than has been possible within the scope of this study. For example, the marginality of the discourse of talent in education policy has been considered only through a literary analysis of the text of policy. However, as texts, policies are not necessarily clear or closed or complete. The texts are the result of compromises at various stages – at the points of initial influence, during the various processes of legislation, the micropolitics of legislative formulation, in the parliamentary process, in the politics and micropolitics of interest group formulations. Only certain influences and agendas are recognized as legitimate at any given point of time. (Ball, 2006, p. 48).

Therefore, the process of policy making needs to be opened up with reference to discussions in parliament, in CABE, in review committee meetings, etc. to arrive at a deeper understanding of how certain issues gain representation in policy texts and others do not. Every policy has 'an interpretational and representational history. It enters a social and institutional context, where the text is received by readers with an equally historical context of response' (ibid.). At any given point, other policies and texts are in circulation which inhibit, influence, or contradict the possibility of enactment of others (ibid). These aspects critically influence the manner a policy constructs a particular way of viewing a category like the talented student.

Additionally, despite its close attention to a government programme, the study has not been able to engage with the boundary between the state and society. Within this research, the perspective of the state has been examined through the documents coming from the two agencies, i.e. the MHRD and the NCERT. The operational understanding of the state has therefore been one of a 'set of government agencies and functions that are clearly marked off from society at large' (Sharma & Gupta, 2006). However, the functions of the state cannot be understood without engaging more deeply with its anthropology and sociology. For example, there is a great need to explore how the beliefs and actions of the officials involved at the various levels of the design, implementation, and assessment of a programme like the NTSE influence the interpretation of policy directives. Similarly, the social perceptions of 'talent' have been explored only from the perspective of scholars. There is a huge requirement to explore how such discourses are constituted in the interface between a programme or policy with media and other aspects of public culture.

There is also a need to investigate at greater depth special populations within the category of the 'talented'. This includes groups who have traditionally had unequal and limited access to education such as girls, children from SC/ST backgrounds, children with disabilities, and those from minority communities like Muslims who tend to be under-represented in the state-selected cohort of the 'talented'. What needs to be explored is how such children develop a philosophical belief about their potential and the kind of support system that they require for high levels of academic achievement.

A question that lingers in my mind is 'Was this national talent search worth the efforts and time that the country has invested in it?' One can say that the NTSE has impacted the lives of a good number of individuals, who have gone on to make their mark at the national and international levels in a variety of disciplines. The NTSE has also definitely not impacted enough numbers of students in post-Independence India. These two aspects therefore cause me to reflect on the legacy of this programme. I have mixed feelings when I think of these 58 years of pursuit of the dream of 'nation-building' through a talent search. Perhaps this is natural because 'nation-building' is a utopian project, where the destination of the perfect society or world keeps ever receding in the face of new challenges. However, the ideas, energies, and the dreams behind such a pursuit may have unexpected outcomes.

After my interview with her, Roshni (1970) sent me some photographs of a small community centre she constructed in Thiruvananthapuram, Kerala. She had initially constructed it using her NSTS scholarship money, most of which she was able to save because she went to a college near her residence. In 2011, after her retirement from her academic career, she renovated this building which was 40 years old at that time. When I met her in 2014, it was flourishing as a centre where a group of children from low-income families came together for evening classes from Monday to Friday. Saturdays were set apart from what Roshni called 'pure fun'. The children's group was called 'Killikoottam', meaning 'a gathering of birds'. While this centre had no special name, it was locally known as the 'Nursery'.…Nursery, nurture, incubation, growth… these associations flooded my mind. Looking at the pictures of this simple centre, it struck me that this too was one sort of a 'nation-building', albeit one which was never anticipated by the designers of the talent search. As a project in perpetual progress and evolution, 'nation-building' does include its share of surprises.

## Note

1  https://www.online-inspire.gov.in/

## References

Ball, S. J. (2006). *Education Policy and Social Class: The Selected Works of Stephen J. Ball*. Routledge.

Burke, E. (2001). *On the Sublime and Beautiful. Vol. XXIV, Part 2. The Harvard Classics*. (The Harvard Classics). Retrieved from Bartleby.com: http://www.bartleby.com/br/02402.html.

Gallagher, J. J. (2000). Unthinkable Thoughts: Education of Gifted Students. *Gifted Child Quarterly*, 44(1): 5–12. doi:10.1177/001698620004400102

Kaiser, Cheryl R. & Major, B. (2006). A Social Psychological Perspective on Perceiving and Reporting Discrimination. *Law & Social Inquiry*, 31(4): 801–830. http://www.jstor.org/stable/4490537

Kumar, K. (2005). *The Political Agenda of Education*. Sage.

Kumar, K. & Sarangapani, P. (2004). History of Quality Debate. *Contemporary Education Dialogue*, 2(1): 30–52.

Naik, J. P. (1982). *The Education Commission and After*. Allied Publishers.

Russell, B. (2010). *Education and the Social Order*. Routledge.

Sharma, A., & Gupta, A. (2006). Introduction: Theoretical Genealogies. In A. Sharma, & A. Gupta, *The Anthropology of the State* (p. 43). Blackwell.

Taylor, C. (2009). *The Politics of Recognition. In Multiculturalism*. Princeton University Press.

# Index

policy: constructions 41, 44–45; design 15; discourses 52, 54; morphology 55–62; texts 21–23, 45–50
pratibhā 33, 34, 185
prestige: benefits of NTSE 121; credibility 130; desire 13; in school 130; preparation 135; science 39, 181, 185; state recognition and 15
Project (NTSE Selection) 73, 80, 82, 112–115, 117, 133, 163, 190

Raman, C. V. 39
Ray, P. C. 39

Saha, M 39
Saxena, K. N. 16, 66
scholarship: NTSE 1, 2, 17, 27, 77–79, 81–82, 109, 120–121, 128–133
scholarship (others) *see* Chacha Nehru Scholarship; Cultural Talent Search Scholarship; INSPIRE; KVPY
schools for the talented: Jnana Prabodhini Prashala 7 *see* Navodaya Vidyalaya
Science Aptitude Test (SAT) 72, 74
science: affective engagement 18, 134, 141; career 79, 127, 132–133, 149, 161–162, 171–172, 175; elite 71; nation-building and 2, 28, 32, 37, 159, 161; popular 39, 141, 161; scientific temper 37
scientist: community 71; heroic 113, 138; influence 38, 39; origins of 66; political influence 81; popularity 141, 185–186; research 159–161; social separation 186; success 163, 181; supply of 81, 79; women 18, 25

social sciences (academic field) 2, 7, 9, 80, 85, 97, 109, 132, 138, 147, 161, 165, 178, 179, 181, 190
strain 76–77
Subject Aptitude Test (SAT) 10
summer school 2, 18, 19, 71, 78, 80, 113, 121, 123–127, 147, 169, 170, 181

talent: adjectives associated with 45–46, 55, 57–60; as a label 3, 8, 13, 15, 52, 53, 56, 58, 82, 105, 133, 139, 177–178, 181, 189; collective nouns associated with 55–60; peers 69, 71, 109, 127, 130, 132, 143, 147, 152, 171, 181; synonyms 22, 45–47, 55, 60–61, 115; usage 30; verbs associated with 55–57; wastage of 63, 147–148, 182
talents: parable 28–29; republican discourse 30
talent search beneficiaries: caste 10, 12, 34, 99, 101, 108, 145, 155–158, 164, 174, 180–182; class 108, 153, 164, 181, 187; gender 10, 12, 16, 23, 24, 52, 98, 102, 103, 134, 140, 148, 149, 150, 178, 182; rural 11, 49, 50, 84, 92–95, 97, 99, 101, 102, 107, 143, 145, 150, 153–154, 156, 158, 164, 171, 173, 175, 178, 180, 183, 184, 190; urban 10, 94–95, 99, 102, 152–154, 178, 180
talent search, Westinghouse 32, 72, 190
teacher, school: discourse of limitations 67–69; guidance in NTSE 112–113; guru 34, 35; of the talented 54, 170; policy 189; rural areas 154–155, 173

World Wars 4, 32, 38, 73, 81, 185

For Product Safety Concerns and Information please contact our EU
representative  GPSR@taylorandfrancis.com
Taylor & Francis Verlag GmbH, Kaufingerstraße 24, 80331 München, Germany

www.ingramcontent.com/pod-product-compliance
Lightning Source LLC
Chambersburg PA
CBHW060300220326
41598CB00027B/4180

* 9 7 8 1 0 3 2 3 8 4 0 8 5 *